BRIDGES AS STRUCTURAL ART

MIGUEL ROSALES

ORO

BRIDGES AS STRUCTURAL ART

MIGUEL ROSALES

To my husband, John David Corey, for his unconditional support and encouragement.

Tilley Bridge, Fort Worth, TX.

CONTENTS

becoming a
BRIDGE DESIGNER

I came to specialize in bridge design through a series of unforeseen events and circumstances. Bridges have not always been my passion, and at the beginning of my professional career I did not expect to become a bridge designer. Over the years I have experienced how a beautiful, innovative bridge that is well integrated into its context can transform a city or region. I feel fortunate that many of my bridges have become symbols of the areas in which they have been built. Here is the story of how I started and who helped and inspired me along the way.

I was born in Guatemala City, Guatemala, in 1961. I was the eldest of four boys. We lived with my parents in a middle-class neighborhood on the outskirts of the city. My father, Marco Antonio Rosales, had been raised in the countryside with modest means. He did well in school and was the first in his family to obtain a university degree, graduating with honors. He attended the public university in Guatemala City in the late 1950s and graduated with a degree in pharmacy. My mother, Lidia Izás, was a dedicated homemaker. They had a good, balanced relationship. My father was cerebral and precise while my mother was artistic and free spirited.

From the start, my education was very important to my parents as they wanted me to achieve professional and personal success. They believed that someday I might go to the United States, and they always talked about the advantages of an advanced degree from a prestigious university abroad. They fostered my dreams and their support and encouragement are still part of my life.

Miguel Rosales at the Longfellow Bridge over the Charles River in Boston, MA, 2019.

Left page, Zakim Bridge in Boston, MA.

My family enrolled me in a private school where I learned English and finished high school. They invested in my education and created an environment to pursue my dreams. My father died when I was twelve, but my mother ensured that I would receive the best education available in Guatemala. Years later I was the valedictorian of my high school graduating class. Although my father was not there to see me give the farewell address, I am sure he would have been proud.

The early encouragement from my parents has had a strong influence on who I am today. I was an intense, methodical, and lonely child always focused on achieving perfection in everything I did. Those personal traits have served me well in my profession, as there are many obstacles to overcome when designing and building an innovative, visually compelling bridge. Through the years I have learned that a combination of vision, patience, and persistence is key.

In 1979, after graduating from Valle Verde High School in Guatemala City, I enrolled at Universidad Francisco Marroquín (UFM), a local private university with high academic standards. UFM is considered the most prestigious university in the country. I graduated six years later, in the spring of 1985, with a professional degree in architecture. I did well in my coursework and graduated with the highest grades in my class. However, although I did excellent work in my other classes, I had difficulty with structural analysis courses. This is interesting since structural analysis is fundamental to bridge design and so important to my professional career today. Back then I did not see the relationship between structural formulas and creative design. How to achieve a beautiful structure or building through rigorous structural calculations was not obvious to me. Later, during my tenure at Boston's Central Artery/Tunnel Project (CA/T) in the 1990s, I would learn that bridge manuals and structural codes were not necessarily written to pursue creative designs. They were mainly developed for the standardization of regulations and implementation of technical and safety requirements. To be a successful bridge designer one needs to know more than how to apply codes and carry out structural analysis. A functional, inexpensive bridge might solve a technical problem. However, unless you consider aesthetics and its surroundings, it will rarely receive community support or become a source of pride as a symbol of an area.

Lidia Izás and her son Miguel Rosales at his First Communion in Guatemala City, Guatemala, 1968.

In 1984 I applied to the School of Architecture and Planning at the Massachusetts Institute of Technology (MIT) to earn a master of science degree. One of my professors at UFM, Adolfo Lau, had attended MIT and encouraged me to apply. He was a successful architect and an effective teacher. I did not apply to less demanding schools because I was striving for the best possible education in the United States.

I went to Boston for the first time in the fall of 1984 where I met with professors and visited the MIT campus. I was impressed with Boston and its architectural heritage, particularly the well-preserved and gracious historic neighborhoods, such as Back Bay and Beacon Hill. I remember how different it felt from Guatemala City. The Charles River and Esplanade Park, along the riverbank, seemed to me to be incredible recreational resources. I never dreamed that in the future I would design the Appleton Bridge in that park and help restore the Longfellow Bridge over the Charles River. By the end of 1984 I had sent in my application and design portfolio to MIT and hoped that my American dream would become a reality.

During the spring of 1985 while I was completing my practical training, finishing my thesis, and preparing for my graduation exams from the School of Architecture at UFM, I was also awaiting a decision from MIT. In early April an admission letter finally arrived. I will never forget the moment I opened the envelope that changed my life forever. The instant I saw the thick MIT envelope I felt elated about my future. I emigrated to the United States in 1985 after completing my degree in architecture from UFM and winning a full fellowship to enroll at MIT from the Organization of American States (OAS). The OAS awarded fellowships to Latin American students for graduate studies at American universities, and I was fortunate to receive one. The fellowship covered everything—tuition, fees, books, housing, and a stipend to live on. I could not have afforded to pay for an MIT education without it. I was the only Guatemalan student in the School of Architecture and Planning and the first student from my country to complete the two-year master of science graduate program. It took substantial effort to adapt to MIT's highly competitive environment, but I found ways to fit in and took advantage of the many opportunities at my disposal by not getting discouraged and applying myself.

Miguel Rosales at his graduation from Universidad Francisco Marroquín in Guatemala, 1985.

Longfellow and Appleton Bridges, Boston, MA.

I completed my MIT degree in June 1987. After graduation I wanted to work on large-scale projects as an urban designer. The Wallace Floyd Design Group, a Boston firm with multiple urban design projects in its portfolio, hired me in 1988. One of my first projects was the preparation of a master plan for a portion of the MIT campus. That area has since been developed as a major gateway into the campus.

My plan was to work as an urban designer for the rest of my professional life. However, my career took a dramatic turn when I started working on the CA/T Project in early 1989. My former employer assigned me to work with a group of urban designers, architects, and landscape architects that was trying to improve the appearance of one of the largest highway transportation projects in the country. I had no experience with the design of bridges or highways because until then I had mainly worked as an urban designer. I also did not know how to fit into a complex engineering organization or contribute to the design of a transformational and challenging transportation project. However, I was excited and ready to start a new chapter of my professional career.

In 1989 the CA/T Project's aesthetics manager was Rebecca Barnes, an architect and planner who worked for the Massachusetts Department of Transportation (formerly the Massachusetts Department of Public Works). She oversaw the architecture, landscape architecture, and urban design departments. Fred Salvucci, Massachusetts secretary of transportation at that time, hired her because he wanted someone to look at the project from a different perspective. One of his goals was to improve the aesthetics of the overall project and the contextual integration of the highway into the city fabric. He was also especially interested in how Interstate 93 would cross the Charles River and connect to a new tunnel under downtown Boston, via a new bridge. Both were pioneers in thinking of highway projects as more than transportation infrastructure. At that time, there was not a strong interest in the aesthetics of highways or bridges, nor was their capacity for visually improving a city like Boston fully understood. Bridge engineers at the CA/T Project were not used to working with architects and designers like myself. The two different ways of thinking and looking at the project often caused friction, but it also created an opportunity for change and innovation, which I took.

When I arrived at the CA/T Project in 1989, I learned that the proposed design for crossing the Charles River was controversial and that environmental agencies and nearby neighborhoods opposed it. The project's engineers had proposed a very wide bridge with many supports in the water. They were mainly trying to find a low-cost solution that addressed the traffic demands and functional requirements, but with complete disregard for the appearance of the bridge and its context. I was intrigued by what this new bridge could become in the future for a city like Boston if aesthetics and urban design were considered in addition to functionality and cost. I quickly became interested in the design of highways and bridges and started to learn what it would take to improve them visually.

The CA/T Project leadership played a critical role in my bridge-design career because they gave me the opportunity to focus on the design of the future bridge over the Charles River, and they supported my ideas. They had seen an early architectural model of a potential cable-supported bridge that I had presented at a meeting, and they were impressed. Intuitively, I thought a cable-stayed bridge would be a good fit for the site and they agreed it was worth exploring this alternative further. Their endorsement and support to proceed with the development of new concepts were key in the process of achieving the bridge design that eventually was built years later.

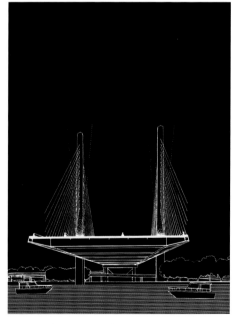

My career as a bridge designer was launched as soon as I was assigned to work on the design of the new bridge. I was quickly promoted and given more responsibility because of my ideas. My path in designing bridges has been long, arduous, and full of obstacles, particularly because I was not initially trained as a bridge engineer. But in the 1990s I was naïve, optimistic, and excited to work on a large river bridge and to eventually see it realized. I did not know that it would take almost fifteen years for my first bridge to be built—the Leonard P. Zakim Bunker Hill Memorial Bridge over the Charles River would not be completed until 2003.

My first step was to prepare a list of bridge design goals from a visual and urban design point of view. I believed that the bridge should cross the entire river with no obstacles in the water, respecting the importance of that body of water in the history of Boston and allowing for unimpeded navigation. I also thought that it would be critical to propose a bridge structural system that would allow for a thin deck configuration since there was limited vertical clearance under the bridge.

Initial cable-stayed bridge concept with two needle towers and ten traffic lanes, 1992.

The bridge had to quickly slope down to enter a tunnel that would replace the elevated highway downtown. I was also convinced that this area, which for many years had been primarily industrial, needed a new identity and that a traditional bridge, like those that for decades had crossed the Charles River, would not help meet this goal. The idea of creating a gateway into the city from the north was appealing and it turned out to be visionary. I proposed a new cable-supported bridge with a long span with no pier supports in the river.

This new bridge type was bold and unique in Boston, and it had never been considered until then. After many studies and preparations of architectural models, I finally arrived at a feasible cable-stayed bridge design in collaboration with the project engineers. This design was immediately well received by the stakeholders as they saw the transformational potential of such an approach. Secretary Salvucci insisted that my proposed bridge solution be included as part of a Supplemental Environmental Impact Statement for the project, which was published in early 1991. The decision had been made that a new cable-stayed bridge structure would be included as part of the project, but many years would pass before the bridge was fully designed and built, as there were still multiple obstacles on the way. The proposed highway configuration and associated interchange next to the crossing, the so-called Scheme Z, was not conducive to the creation of an elegant and beautiful cable-stayed bridge. There were too many ramps connecting to the bridge at different elevations and the bridge width was not constant over the water. However, I was optimistic about the future of the bridge and my role as its designer and architect.

In the summer of 1991, newly elected Massachusetts governor Bill Weld created a Charles River Crossing Review Committee to improve the appearance of the interchange Scheme Z and the river crossing. This group included representatives from several environmental and regulatory agencies in addition to neighborhood groups and other professional organizations. Without their participation, it would not have been possible to modify the interchange and reduce the number of aboveground lanes crossing the river to achieve a better bridge design. We met for almost two years as part of a process that I was assigned to assist. The committee decided that the CA/T Project should hire a bridge expert to work directly for them as an independent advisor. After a comprehensive search, they retained Christian Menn, a well-known Swiss bridge engineer whom I had recommended to the committee.

Miguel Rosales and Christian Menn at the Central Artery/Tunnel Project in Boston, MA, 1993.

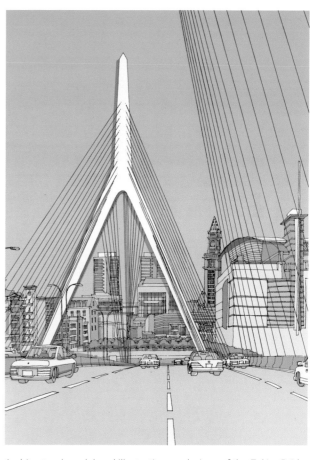

Architectural model and illustrative renderings of the Zakim Bridge prepared during the project's final design phase, 1994.

Next page, illustration showing the Zakim Bridge in the context of the cities of Boston and Cambridge, MA, 1994.

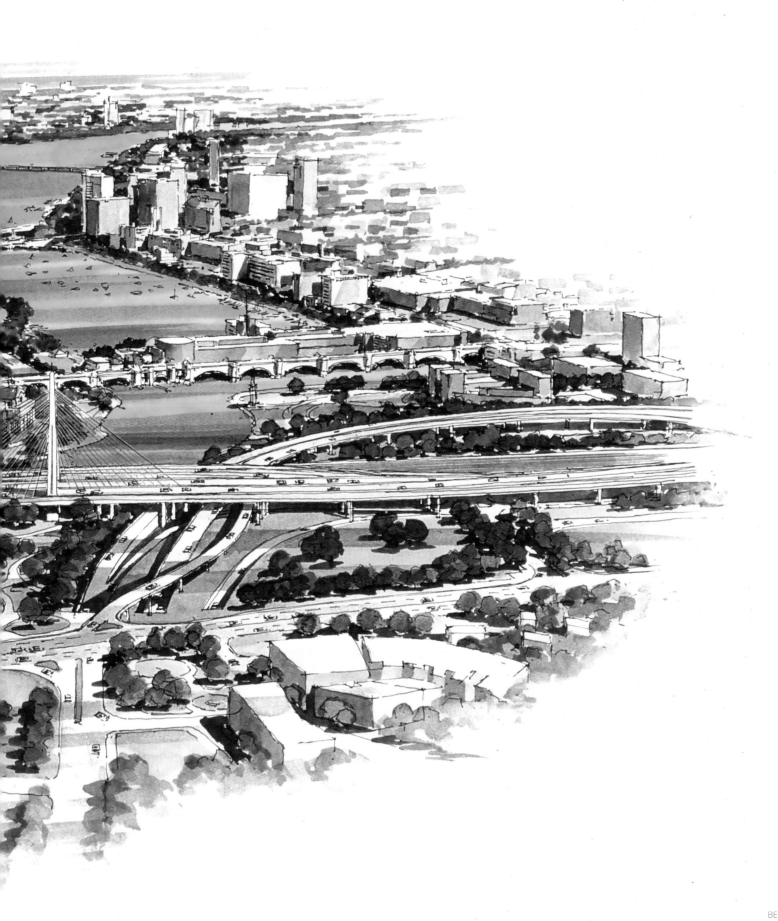

I had previously heard Christian give a lecture at Harvard University and was impressed by his credentials and the overall quality of his bridges. I remember meeting Christian for the first time at the CA/T Project offices in late 1991. I showed him the architectural model of the cable-stayed bridge I had proposed for the Charles River crossing. He immediately said that the bridge concept was appropriate for that location and that he would endorse my approach. His expert opinion validated my early decision and bridge design proposal. I was eager to work with him and explore different options. He became my mentor and teacher.

By 1993, after multiple meetings with the Charles River Crossing Review Committee and project officials, the original Scheme Z highway proposal was finally abandoned, and the width of the river crossing was substantially reduced and straightened, which was important to achieve an elegant bridge. New vehicular tunnels were also added in the vicinity of the bridge, allowing several elevated ramps for the adjacent interchange to be reduced and lowered. A new cable-stayed bridge concept had also been conceived through a long process guided by Christian Menn, similar to the bridge that was eventually completed in 2003. I continued to work on the project, refining and detailing the Zakim Bridge design until 1996, always in consultation with Christian Menn, who had returned to Switzerland. Working directly with Christian was a great privilege in the early stages of my career and this experience created an excellent foundation for my professional development.

After completing my tenure at the CA/T Project in 1996, I realized that I would need to continue my education and learn more about bridge engineering to pursue a career in bridge design on my own. I applied for and won several grants and scholarships to pursue further studies in bridge design and engineering. I traveled to Europe to visit famous bridges and study them carefully in addition to spending a summer researching bridges at the ETH in Zurich, Switzerland. I also took engineering courses at Northeastern University in Boston. After visiting and analyzing well-known bridges in multiple countries, I realized that three critical aspects should be considered when designing a bridge: transparency and slenderness, simplicity and structural order, and artistic shaping. My bridge designs always try to achieve these qualities. I am also committed to achieving solutions that are cost-effective and buildable in the United States. Seeing my vision realized and completing a bridge that reflects each of these qualities is always a highlight of my professional life.

Miguel Rosales at the Zakim Bridge construction site in Boston, MA, 2001.

Left, Zakim Bridge during construction, to the right of the bridge it will replace.

By late 1996 the Zakim Bridge plans were almost completed and construction began in full in 1998. I was now ready to start on new bridges after close to seven years at the CA/T Project. I wanted to create my own firm. At about the same time I was fortunate to meet Jörg Schlaich in Boston, a well-known structural engineer who had designed innovative pedestrian bridges in Europe. He was a consultant during the final design of the Zakim Bridge, and Christian Menn introduced me to him as they knew each other well. We collaborated for a few years on bridge projects, and I visited him frequently in Germany, where his office was located. He was an important mentor at the beginning of my career, and I am grateful for his friendship, guidance, and generosity.

The Zakim Bridge was completed in 2003, almost fifteen years after I had started working at the CA/T Project. The bridge was initially named after respected civil rights activist Leonard P. Zakim. He was the former New England director of the Anti-Defamation League and founder of the Lenny Zakim Fund to fight poverty and racism in Boston. I was pleasantly surprised when the name of the bridge was announced because of its symbolism—Mr. Zakim was committed to "building bridges" between people of different races, backgrounds, and religions. After the name was unveiled, I learned that neighborhood leaders in Charlestown, next to the bridge, thought it should be named after a local hero born in Boston. In the spirit of compromise, the Massachusetts legislature added Bunker Hill to the name, to celebrate the American colonists who fought the British in the Battle of Bunker Hill. The shape of the top of the bridge tower was inspired by the Bunker Hill Monument, which is visible from the bridge and ties the name to a local historical event. Over the years I stayed in contact with the Zakim family and in 2018 we jointly celebrated the fifteenth anniversary of the bridge opening. The official name is the Leonard P. Zakim Bunker Hill Memorial Bridge, but nowadays everybody calls it the Zakim Bridge.

I started Rosales + Partners in 1997 (originally Rosales Gottemoeller and Associates) and for the next twenty-five years designed and built bridges across the United States and abroad. I became sole owner of the firm in 2004. My office includes staff with expertise in bridge architecture, structural engineering, landscape architecture, and aesthetic lighting, all critically important and interrelated disciplines required for the conception of unique and elegant bridges.

Jörg Schlaich with Miguel Rosales in Berlin, Germany, 2010.

Right page, Liberty Bridge, Greenville, SC.

In the late 1990s, right after my company was formed, we secured two important bridge design commissions. First, we won the contract for the design of the Woodrow Wilson Bridge, as part of a large consultant team, via a national bridge design competition. This bridge crosses the Potomac River near Washington, DC. It was completed in 2008. The second bridge, located at Falls Park in Greenville, South Carolina, crosses the Reedy River. When it was completed in 2004, the Liberty Bridge became an instant symbol of the city. Both bridges were instrumental in getting early national recognition and establishing a reputation for bridge design excellence that helped secure the future of the firm.

In 2005 Adolfo Lau, my former professor at Universidad Francisco Marroquín in Guatemala, visited me in Boston. I took him to see the recently completed Zakim Bridge as I was proud of my achievement after many years of hard work. I am still grateful that he helped me get accepted at MIT. His encouragement and belief that I could succeed motivated me to apply to MIT, move to Boston, and start a new life. With the completion of important bridges near and across the Charles River, including the Zakim, Appleton, Longfellow, and soon-to-be-completed Charlestown Bridge, I have left my mark on Boston. I see those bridges as a gift to my adopted city and its residents.

During my career I have concentrated solely on bridge design. I felt it was important to specialize to achieve the best possible solutions in the field. When I start a new bridge project, I always consider structural suitability, overall appearance, and costs. I take inspiration from the location and overall context; the setting, site topography, and landscape always inform my decisions. Many of my bridges have become icons in the cities and regions in which they have been built, winning multiple design and engineering achievement awards along the way, and, more importantly, becoming sources of community pride. My American dream has become a reality.

This book showcases bridges in which I have been involved as a bridge designer and architect. I am grateful to have had the opportunity to work on these projects and appreciate the confidence and support from many clients and public officials in the cities and regions in which I have practiced. Their willingness to hire me, support my ideas, and take calculated risks has allowed me to flourish as a bridge designer. It is my hope that readers will find these bridges thought provoking, inspiring, and true examples of bridges as structural art.

Adolfo Lau with Miguel Rosales at the Zakim Bridge over the Charles River in Boston, MA, 2005.

Left page, Aerial view of Charles River bridges, Boston, MA.

BRIDGE
locations

1 Zakim Bridge
Boston, Massachusetts

2 Charlestown Bridge
Boston, Massachusetts

3 Longfellow Bridge
Boston, Massachusetts

4 Appleton Bridge
Boston, Massachusetts

5 Northern Avenue Bridge
Boston, Massachusetts

6 Markey Bridge
Revere, Massachusetts

7 Fore River Bridge
Quincy, Massachusetts

8 East 54th Street Bridge
New York, New York

9 Woodrow Wilson Bridge
Washington, DC

10 Columbus Airport Bridges
Columbus, Ohio

11 Hickory Riverwalk Bridge
Hickory, North Carolina

12 Liberty Bridge
Greenville, South Carolina

13 Throop Street Bridge
Chicago, Illinois

14 River North Bridge
Nashville, Tennessee

15 I-74 Mississippi River Bridge
Quad Cities, Iowa and Illinois

16 Huntsville Bridge
Huntsville, Alabama

17 Griffin Bridge
Des Moines, Iowa

18 Carver Bridge
Des Moines, Iowa

19 Bruce Vento Bridge
Saint Paul, Minnesota

20 Como Park Bridge
Saint Paul, Minnesota

21 Tilley Bridge
Fort Worth, Texas

22 Panther Island Bridges
Fort Worth, Texas

23 Moody Bridge
Austin, Texas

24 Marion Street Bridge
Seattle, Washington

25 Puente Centenario
Panama Canal, Panama

Como Park Bridge Saint Paul, MN

Markey Bridge Revere, MA

East 54th Street Bridge New York, NY

Marion Street Bridge Seattle, WA

Appleton Bridge Boston, MA

Griffin Bridge Des Moines, IA

Moody Bridge Austin, TX

Tilley Bridge Fort Worth, TX

100 ft. (30 m)

PEDESTRIAN BRIDGES
comparison

Liberty Bridge Greenville, SC

Bruce Vento Bridge Saint Paul, MN

Hickory Riverwalk Bridge Hickory, NC

Northern Avenue Bridge Boston, MA

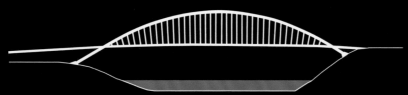

River North Bridge Nashville, TN

Huntsville Bridge Huntsville, AL

250 ft. (76 m)

100 ft. (30 m)

100 ft. (30 m)

Columbus Airport Bridges Columbus, OH

Carver Bridge Des Moines, IA

Throop Street Bridge Chicago, IL

Panther Island Bridges Fort Worth, TX

I-74 Mississippi River Bridge Quad Cities, IA and IL

Charlestown Bridge Boston, MA

500 ft. (152 m)

VEHICULAR BRIDGES
comparison

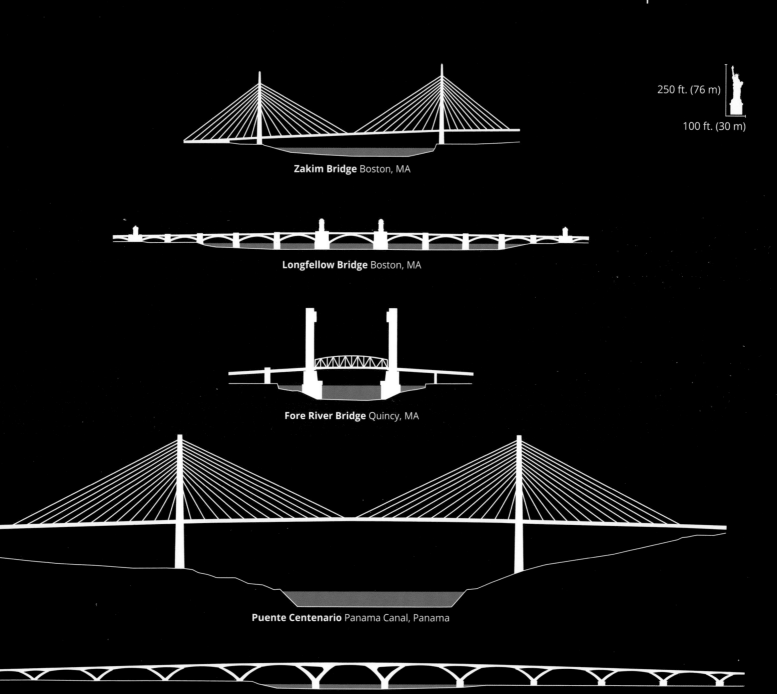

250 ft. (76 m)

100 ft. (30 m)

Zakim Bridge Boston, MA

Longfellow Bridge Boston, MA

Fore River Bridge Quincy, MA

Puente Centenario Panama Canal, Panama

Woodrow Wilson Bridge Washington, DC

500 ft. (152 m)

ZAKIM
bridge

ZAKIM
bridge

Location
Boston, MA

Completion date
2003

Width
183' (55.7 m)

Total length
1,457' (444 m)

Main span length
745' (227 m)

Height of towers
322' (98.1 m) north
295' (89.9 m) south

Navigation clearance
40' (12.1 m) average

The Leonard P. Zakim Bunker Hill Memorial Bridge over the Charles River acts as a gateway into Boston, Massachusetts, from the north. The bridge was completed as part of the Central Artery/Tunnel Project that replaced an elevated highway in the downtown core with a tunnel and parks above it.

The striking cable-stayed bridge architectural form was born out of a multitude of functional requirements and stringent site constraints. The bridge has an unusual configuration; the main span over the Charles River navigation channel is fabricated in steel, and the back spans on land are constructed of heavy concrete to help balance the structural loads. The bridge's unique geometry has cables over the river anchoring at the edges of the deck, and cables over the land anchoring at the median sections of the bridge. This cable configuration creates a memorable, dynamic experience when crossing the river as the angles of the cables are constantly changing. The upper section of the bridge's inverted Y-shaped towers was designed to resemble the adjacent Bunker Hill Monument. The overall configuration creates a structure with a high degree of three dimensionality and visual appeal. The bridge's form echoes the shape of sailboats that navigate the Charles River, giving a new image to a former industrial area along the river. Since its completion, the areas surrounding the bridge have experienced substantial development. New parks and public amenities near the bridge have also opened. The Zakim Bridge is now a contemporary symbol of the city.

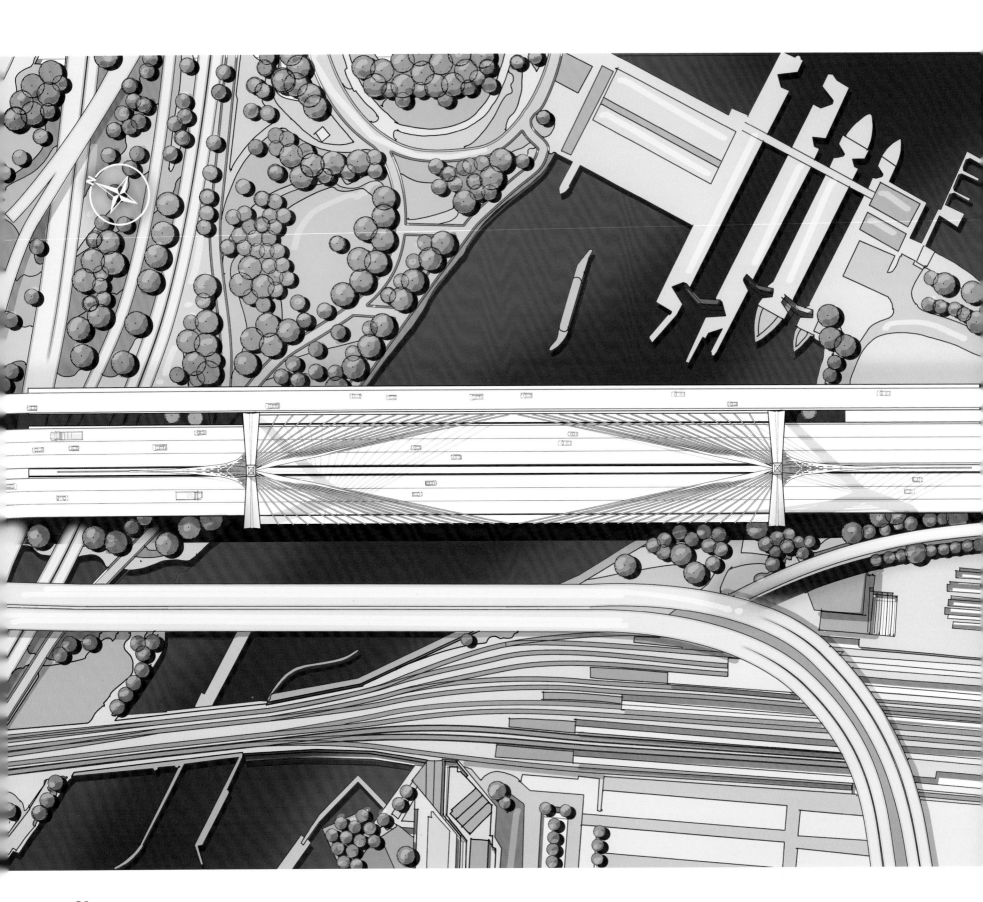

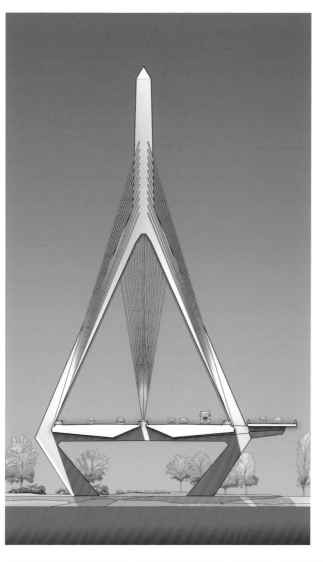

Design innovations and technological advancements at the time of completion:

- First cable-stayed bridge in New England.

- First asymmetrical cable-stayed bridge in the United States, with eight lanes between tower legs and a two-lane cantilevered ramp on the harbor side.

- First-time use of a hybrid configuration. Steel superstructure for main navigational span. Concrete superstructure for back spans.

- Exceptional deck width—widest cable-stayed bridge worldwide at the time of completion.

- Use of cable cross ties, cable pipe surface modifications, and external dampers.

- Unique cable configuration that creates a variety of changing views with back span cables anchored at the medians.

Bridge architectural features and detailing:

- Concrete towers with faceted surfaces that help the bridge appear slender from a distance due to the light/shadow effect created by the tower cross section.

- Bridge deck openings in the median and in the space between the eight-lane main roadway and the two-lane cantilevered ramp help bring sunlight to the water, mitigating shadow impacts for fish migration into the Charles River from Boston Harbor.

- Cable anchors were designed to clearly show how the cables attach to the deck, emphasizing their cylindrical shape and function.

- Shape of the upper portion of the towers inspired by the adjacent Bunker Hill Monument.

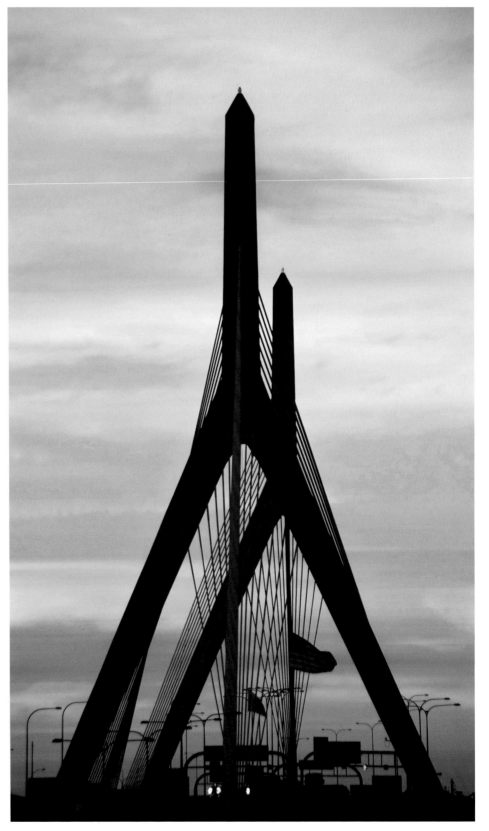

CHARLESTOWN
bridge

CHARLESTOWN
bridge

Location
Boston, MA

Estimated completion date
2025

Width
100' (30.4 m)

Total length
1,088' (331.6 m)

Main span length
250' (76.2 m)

The new Charlestown Bridge, also known as the North Washington Street Bridge, between Boston's historic Charlestown and North End neighborhoods, will replace a century-old truss bridge located at the Charles River delta. The bridge carries two vehicle lanes in each direction, an inbound dedicated bus lane, separated bicycle paths, and generous sidewalks. In addition, the bridge includes two curved overlook areas with granite seating, trees and plants, a shade trellis, and interpretative signage showcasing Boston's historical landmarks visible from the bridge. The use of landscaping to separate and buffer transportation modes will be unique in the city of Boston.

The bridge is part of the Freedom Trail, a pedestrian trail marked in red bricks that meanders through Boston's historic neighborhoods telling the story of the American Revolution by guiding visitors to several historical landmarks in the city center and across the river. About 40 percent of the width of the bridge will be dedicated to sustainable modes of transportation, making it a truly multimodal facility. The crossing will also improve water flow and conditions for boaters by reducing the number of piers from twelve to only five, allowing for a wider navigation channel with a new protective fendering system. The design of the five groups of Y-shaped piers and overall bridge architecture is inspired by the adjacent Zakim Bridge as the two bridges will be seen together from Boston Harbor. When completed, the Charlestown Bridge will become a recognizable landmark in the city due to its scale and visual appeal.

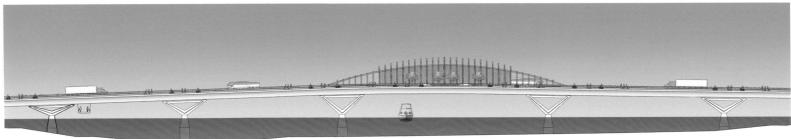

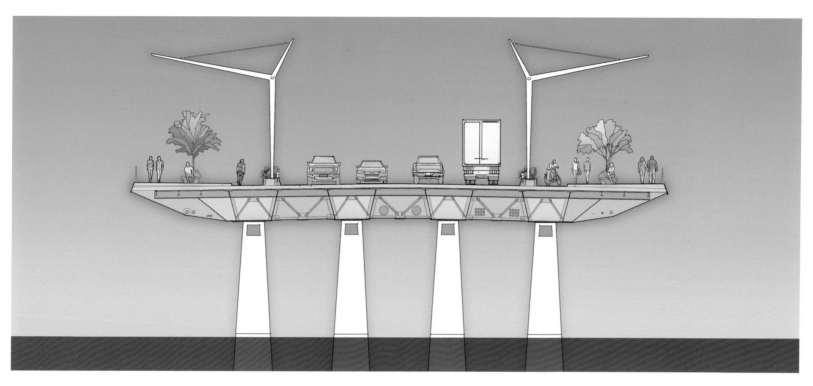

Bridge architectural features and detailing:

- Y-shaped piers are visually compatible with the cable-stayed towers of the nearby Zakim Bridge.

- Use of native plants and trees to separate and buffer transportation modes.

- Light-gray color that enhances the marine coastal environment.

- Architectural arched trellis to mark the main navigation channel from a distance.

- Aesthetic lighting to enhance the bridge at night with changing color capabilities for special events.

- Custom-designed granite benches, planters, and interpretive signage that highlights the city's historical and contemporary landmarks visible from bridge overlooks.

Left, Charlestown Bridge construction, 2023.

LONGFELLOW
bridge

LONGFELLOW
bridge

Location
Boston, MA

Completion date
2019

Width
105' (32 m)

Total length
2,132' (649.8 m)

Main span length
189' (57.6 m)

Navigation clearance
26' (7.9 m)

The landmark Longfellow Bridge was completed in 1906. It is the most important historic bridge in the Boston area due to its prominent location over the Charles River as well as its visual quality and historic integrity. Originally called the Cambridge Bridge, in 1927 it was renamed the Henry Wadsworth Longfellow Bridge in honor of the distinguished local poet.

The multimodal bridge carries MBTA trains, motor vehicles, pedestrians, and bicyclists over the Charles River. The bridge substructure is built of granite masonry and consists of ten hollow piers and two abutments. The central bridge arch span is marked by four neoclassical granite towers, which are the origin of its popular nickname the Salt and Pepper Bridge.

The primary goals of the rehabilitation work were to address the bridge's structural deficiencies, upgrade its capacity, and bring it up to date with modern codes while preserving its architectural character. The original steel arches were preserved and refurbished. The bridge's ornate cast-iron pedestrian railings were restored or replicated when missing. Its granite masonry was cleaned and conserved. A new lighting system was installed to highlight the arches and towers at night with blue lighting. Original light posts and lanterns were replicated and placed along the bridge sidewalks, which were widened. Dedicated, protected bicycle lanes were installed. The Longfellow Bridge has once again become a source of pride for the community due to its enhanced artistic merit and the careful restoration and preservation of its historical character and detailing.

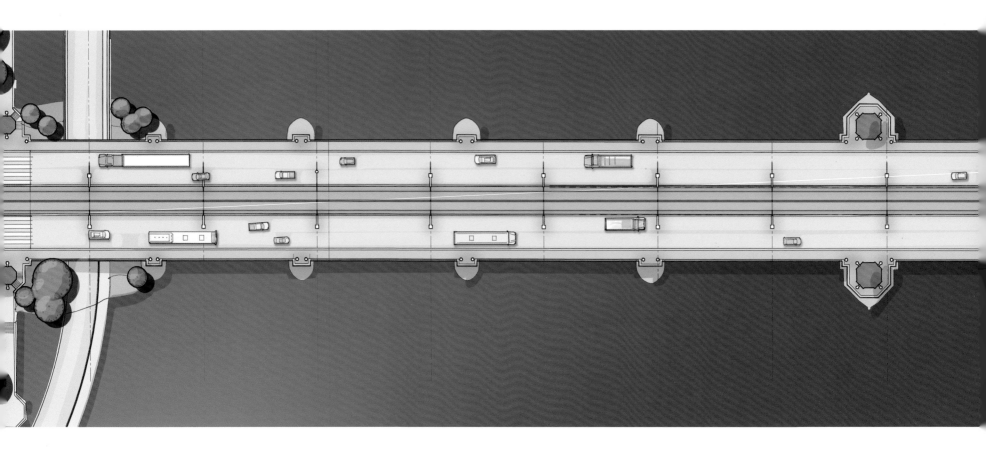

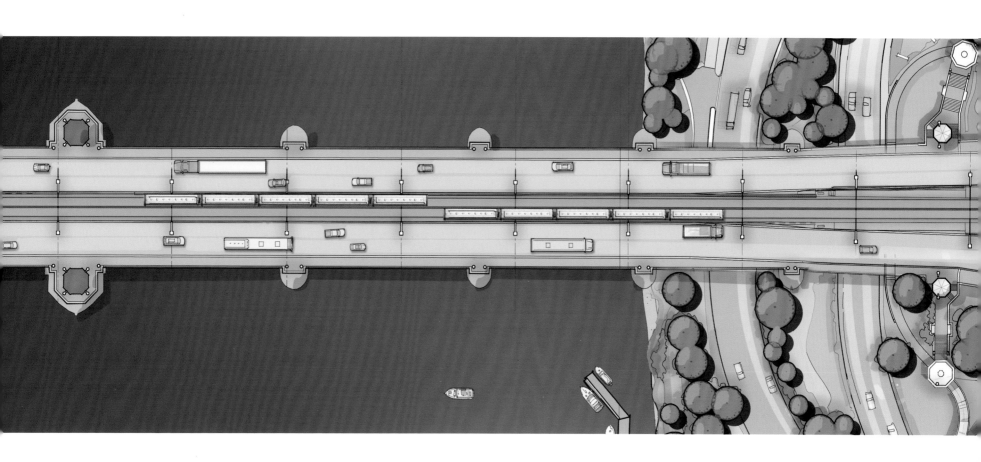

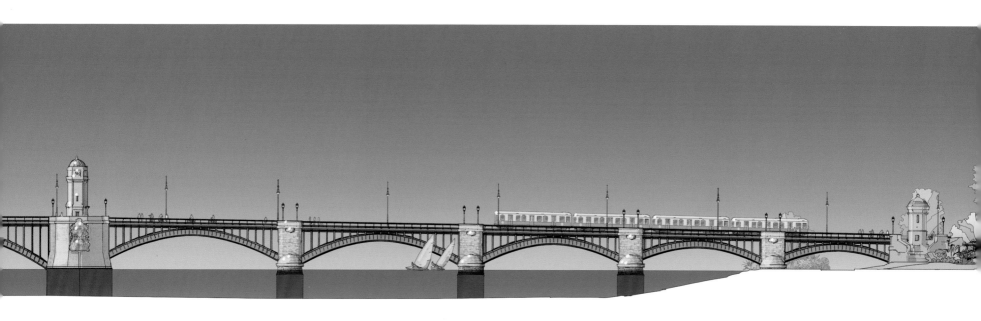

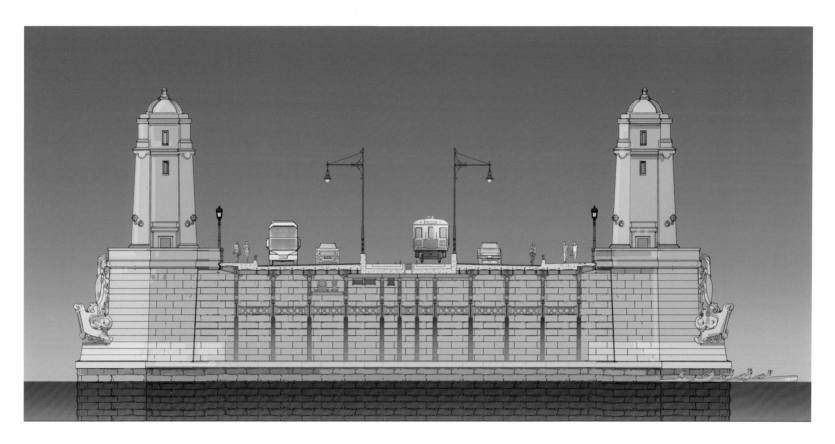

Historical character-defining bridge elements that were saved, restored, and enhanced:

- Steel arches and pedestrian cast-iron railings were repaired, restored, and coated.

- Four main granite towers and four smaller abutment granite towers were cleaned, restored, and structurally strengthened.

- New illumination of steel arches; replicas of tower lighting, streetlight poles, and luminaires; and new interior lighting of four main granite towers were installed and integrated into the overall structure.

- Bronze doors, decorative grates, and wood windows were restored or replicated and installed at the granite towers.

- A new path system and landscaping were added to reconnect the bridge to its parkland setting, via its original monumental granite stairs.

Before bridge restoration and rehabilitation, 1995.

APPLETON
bridge

APPLETON
bridge

Location
Boston, MA

Completion date
2018

Width
14' (4.2 m)

Total length
607' (185 m)

Main span length
222' (67.6 m)

Vehicular clearance
14.5' (4.4 m)

The Frances Appleton Bridge is a multiuse structure on the banks of the Charles River in Boston, Massachusetts. It connects the historic Beacon Hill neighborhood to the riverfront Esplanade. The arch bridge has a clear span over Storrow Drive, a busy arterial roadway that separates the city from the river.

The bridge alignment includes the use of generous curves to make the journey across more enjoyable for pedestrians and bicyclists by providing a variety of changing views. The bridge achieves visual transparency and lightness due to its continuous steel superstructure, consisting of slender steel box girders, which are curved in two directions and branch into two curved staircases that frame a scenic overlook with river views. The bridge features an architectural steel fascia that has a constant depth for visual consistency. The main steel arch was designed to complement the adjacent arches of the historic Longfellow Bridge but with a contemporary appearance. The shape of the arch—wider at the crown and narrower at the abutments—minimizes impacts on the park. The bridge approaches include Y-shaped steel piers intended to visually resemble branches of existing mature trees in the park.

The bridge is named after Frances Appleton, wife of Henry Wadsworth Longfellow, to celebrate their union. Both bridges are visually compatible because of their similar structural arch systems, matching gray-green color, and overall visual quality. The Appleton Bridge is characterized by its technical innovation, elegant detailing, and context-sensitive design, which integrates well into the parkland and historic riverfront.

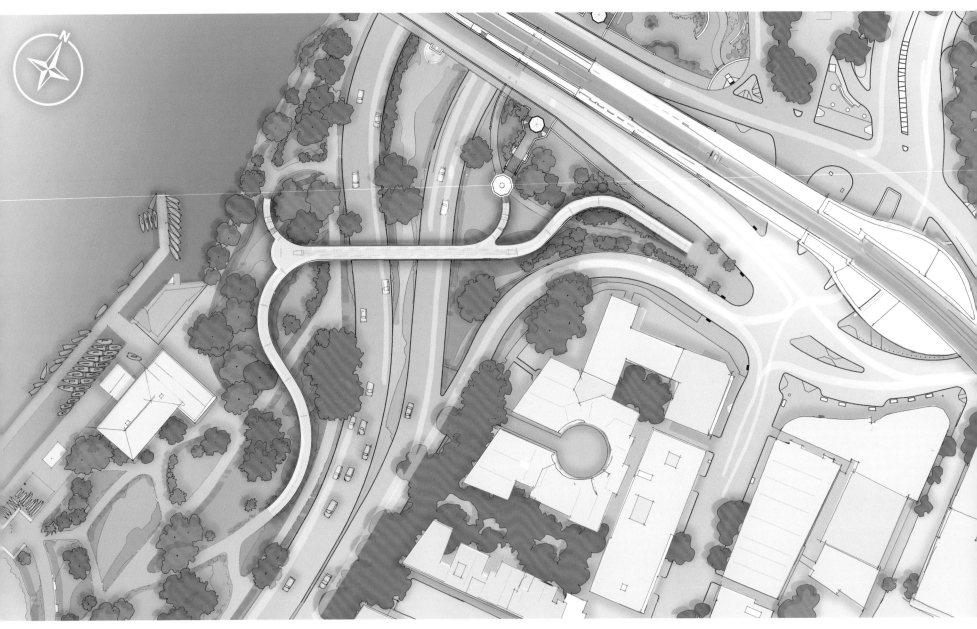

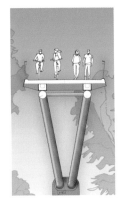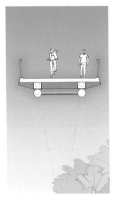

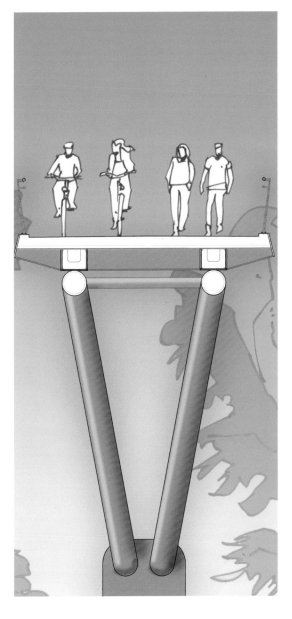

Innovations and technological advancements at the time of completion:

- First accessible pedestrian/bicycle bridge over Storrow Drive, which connects Boston to its riverfront Esplanade.

- Use of steel castings to accommodate special detailing of Y-shaped piers to resemble tree branches.

- Use of lightweight concrete to help reduce pedestrian-induced vibrations and avoid dampers.

- Pedestrian bridge with the longest arch span in the Boston area.

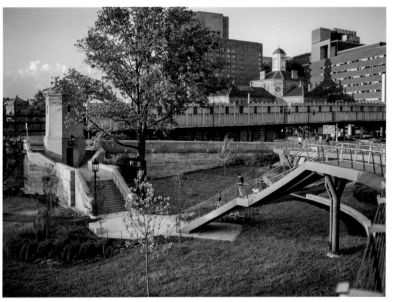

NORTHERN AVENUE
bridge

NORTHERN AVENUE
bridge

Location
Boston, MA

**Estimated
completion date**
(to be determined)

Width
92' (28 m)

Total length
615' (187.4 m)

Height of tower
170' (51.8 m)

Navigation clearance
16' (4.8 m)

The Northern Avenue Bridge will replace a movable truss swing bridge that spans the Fort Point Channel along Boston's waterfront. The historic swing bridge was completed in 1908 and was in service for over a century before it was closed in 2014 due to its advanced state of disrepair.

The replacement bridge is inspired by the historic bridge with its central tower in the middle of the channel at the same place that the movable bridge machinery was originally located. The fixed bridge is a single-tower suspension bridge with two equal side spans that symbolically relate to the long truss cantilevers of the old movable swing bridge. The cables are designed in an angular zigzag arrangement that recalls the truss structural system of the historic bridge. The slender conical towers are reminiscent of tall ships and sailboats in the harbor. The bridge will accommodate pedestrians, bicyclists, and trolleys as they travel between the Financial and Seaport districts. Bicyclists and trolleys will be located between the towers while pedestrians will be on the edges of the bridge on wide sidewalks. At the towers, the sidewalks have been widened to create areas for seating and landscaping where pedestrians can rest and admire harbor views. The shape of the bridge, with its elegant catenary cables, is visually compatible with the marine setting. The Northern Avenue Bridge will become the first long-span suspension bridge in the city and a new waterfront landmark.

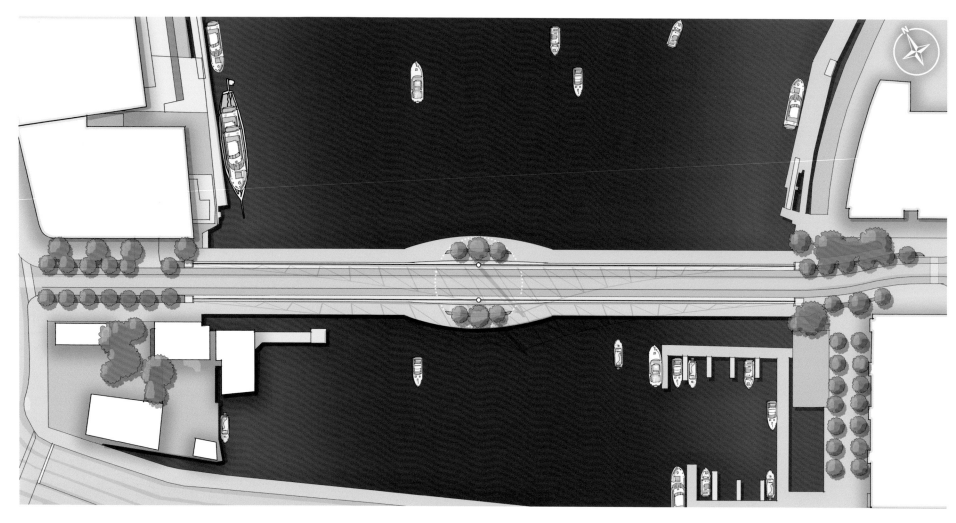

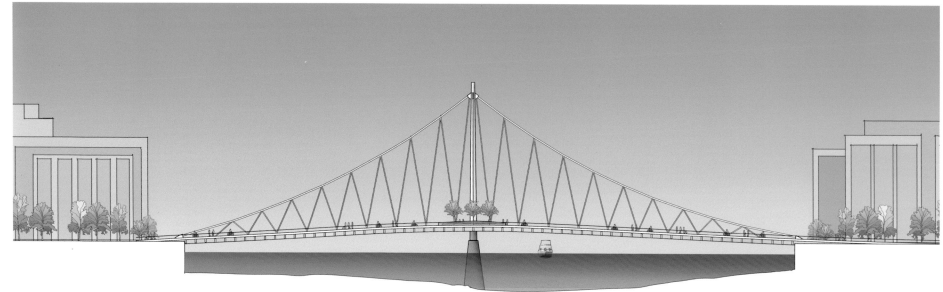

Bridge architectural features and detailing:

- First single-tower contemporary suspension bridge in the region.

- Use of inclined stainless steel railings with horizontal cables to create open views of the waterfront.

- Tapered vertical steel towers inspired by the masts of traditional ships in the harbor.

- Catenary cable system that complements the marine environment.

- Aesthetic lighting enhances the bridge at night.

MARKEY
bridge

MARKEY
bridge

Location
Revere, MA

Completion date
2013

Width
12' (3.6 m)

Total length
151' (46 m)

Main span length
107' (32.6 m)

Height of tower
52' (15.8 m)

The Christina and John Markey Memorial Bridge is in Revere, Massachusetts, adjacent to the Atlantic Ocean. It was built as part of a transit facility project that connects the Wonderland MBTA station to Revere Beach over Ocean Boulevard. Revere Beach, dedicated in 1896, is the first public beach in the United States. The bridge has helped popularize the beach once again after a period of decline that started in the 1970s. New hotels, residential buildings, and restaurants have been built near the bridge to take advantage of the open views of the ocean, framed by the elegant structure. The bridge is named after the late parents of Edward Markey, United States senator from Massachusetts who helped secure funding for the project.

A pair of outward-inclined steel towers with a V configuration frame the access to the ocean, providing enhanced views of the waterfront. The tapered rectangular towers are visible from a distance and mark the crossing location of a pair of historic beach pavilions. The bridge can be accessed via an elevator and stairs that are located at street level. It also has direct access via a plaza that links the bridge to the subway station, which was improved at the same time. The crossing has six front stays and six back stays for each mast. The main span over the boulevard is longer and the back span over the land is shorter. The back stays are closer together to emphasize the asymmetry of the bridge, which connects to a curved plaza landscaped with native beach plantings. Aesthetic lighting enhances the night experience for all users. The Markey Bridge has given a new identity to the city of Revere and become a catalyst for redevelopment in the area.

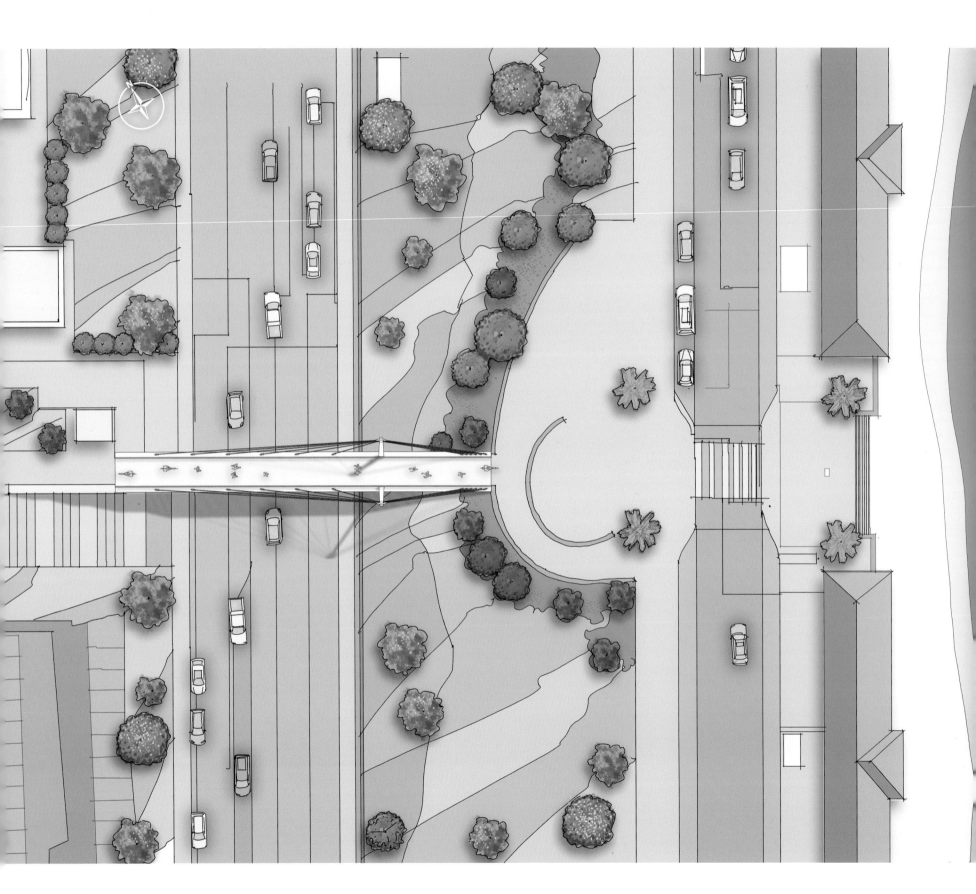

Bridge architectural features and detailing:

- Use of inclined stainless steel railing with horizontal cables to create open views of the waterfront.

- Tapered inclined steel towers; thin steel longitudinal girders; and well-detailed cable connections.

- Light-gray color complements the marine beach environment.

- Tapered abutments and tower base with horizontal rustications for aesthetic appeal.

- Aesthetic lighting enhances the bridge at night with changing color capabilities for special events.

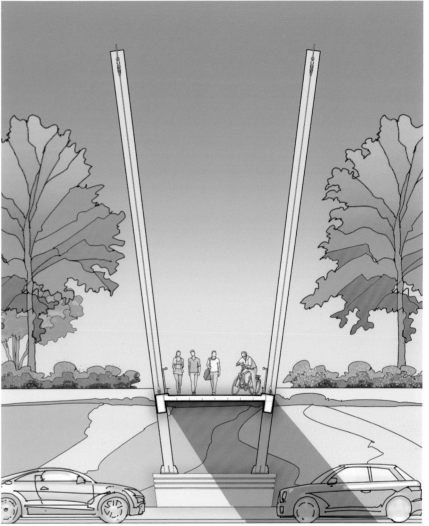

FORE RIVER
bridge

FORE RIVER
bridge

Location
Quincy, MA

Completion date
2018

Width
79' (24 m)

Total length
2,406' (733.3 m)

Main span length
324' (98.7 m)

Height of towers
275' (83.8 m)

Vertical clearance
closed position 60' (18.2 m)
open position 220' (67 m)

Horizontal navigation
channel clearance 250' (76.2 m)

The Fore River Bridge is a vertical lift bridge that carries Route 3A over the river between Quincy and Weymouth, Massachusetts. The Fore River is a vital waterway that provides navigable access to major industries, including oil and gas distribution terminals that serve coastal Massachusetts. The bridge replaced a 1936 bascule bridge that was demolished in 2004 due to its advanced state of disrepair.

The vertical lift span accommodates Panamax-size tankers and other large vessels used by local industries, which will extend the life of the bridge's functionality for decades to come. The new bridge improved bicycle and pedestrian safety and expanded the previous horizontal and vertical channel clearances. A lower frequency of bridge openings and improved traffic performance along the busy corridor, especially during the summer months, were important goals achieved. Each tower foundation consists of two reinforced, concrete plinth structures, which were enhanced architecturally to resemble traditional granite foundations. The towers and the machinery within them are clad with a stainless steel mesh. This forms a semitransparent skin that reflects light during day and night, and highlights the towers from a distance while concealing lift equipment, stairs, and elevators. The architectural detailing of the railings, tower facades, and roadway lighting is visually consistent along the entire bridge corridor and inspired by the art deco style of the historic bridge. The Fore River Bridge is a new landmark in the area that contributes to the region's economic vitality and visually enhances the riverfront.

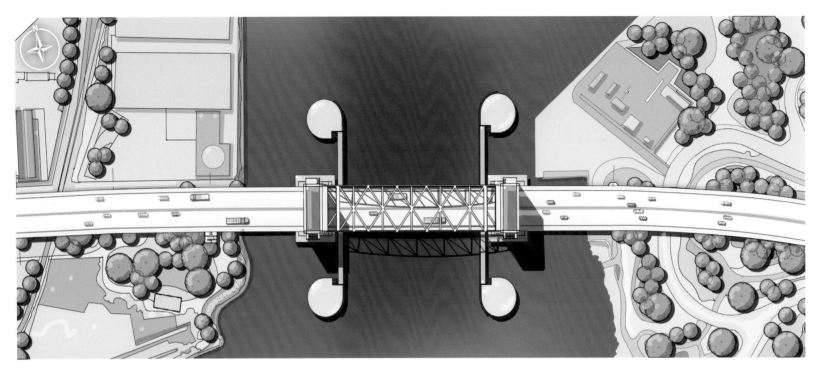

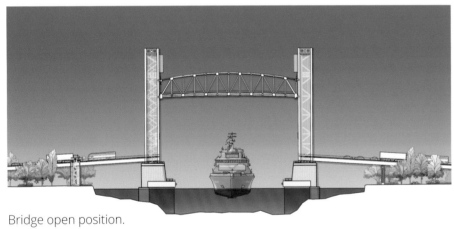

Bridge open position.

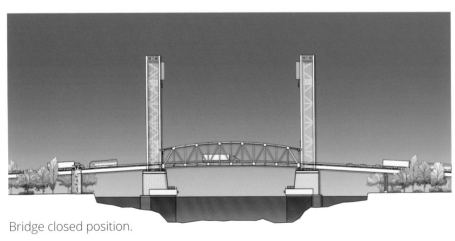

Bridge closed position.

Bridge architectural features and detailing:

- Use of stainless steel semitransparent mesh enclosure for towers and control center helps to conceal utilities, stairs, and vertical mechanical circulation to enhance tower facades.

- Contemporary art deco inspired detailing and articulation of safety railings, roadway light poles, concrete piers, and steel towers.

- Aesthetic blue lighting enhances the bridge at night.

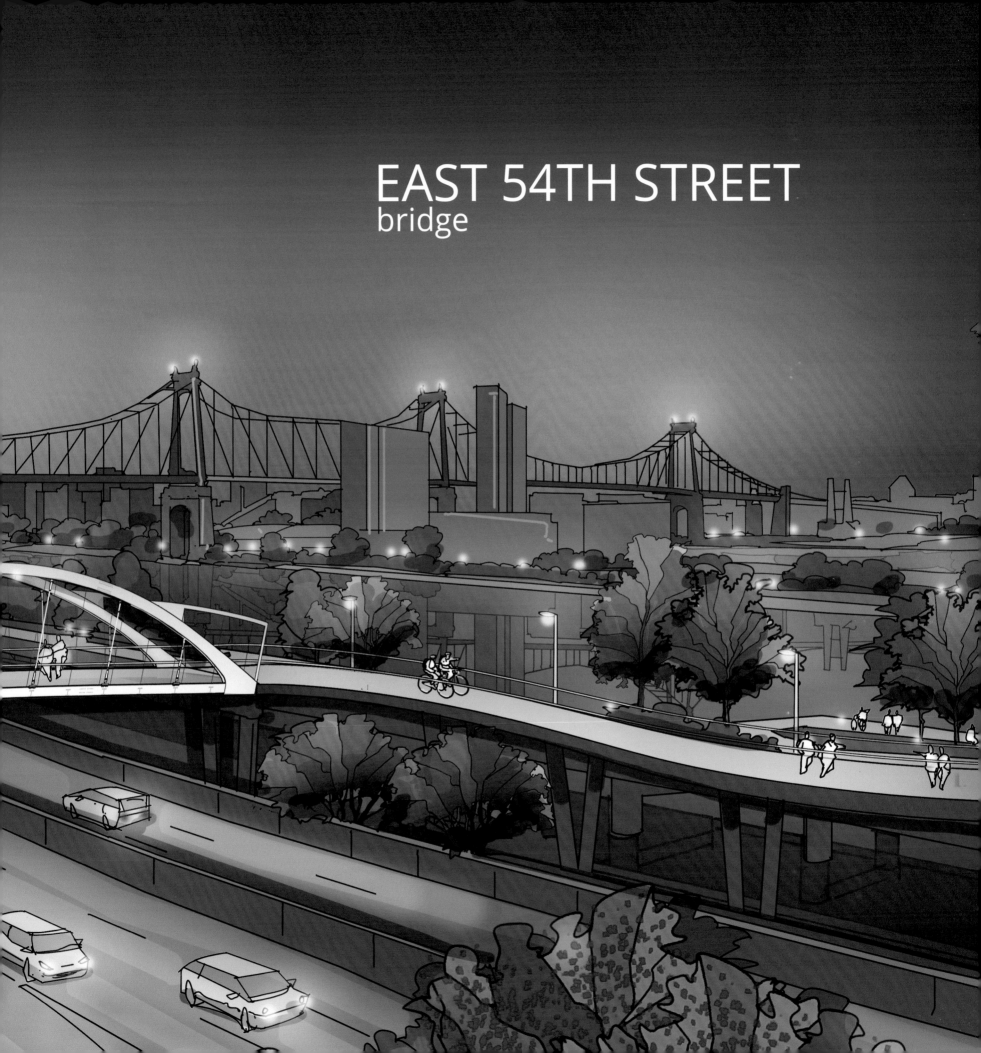

EAST 54TH STREET
bridge

EAST 54TH STREET
bridge

Location
New York, NY

Completion date
2024

Width
14' (4.2 m)

Total length
272' (82.9 m)

Main span length
116' (35.3 m)

Height of arch
18' (5.4 m)

Vehicular clearance
16' (4.8 m)

The East 54th Street footbridge crosses over Franklin D. Roosevelt East River Drive (FDR Drive) in New York City's Upper East Side. It connects with a new linear park and walkway, located on the East River as part of the East Midtown Greenway waterfront project. This project closes the loop of a long-planned New York City landscaped link between East 30th and 61st Streets. FDR Drive was constructed in the 1930s in a narrow corridor between the city and the river, creating a physical barrier for decades. The arch bridge links to the new park at 54th Street next to the Sutton Place residential neighborhood. The bridge was built without major disruptions to the flow of vehicular traffic during construction because it was fabricated off-site and brought over by river barge.

The bridge arches are tilted outward with minimum bracing to create a more open and unique experience. The pedestrian railings and ramp support piers are also inclined at the same angle to reinforce the bridge's overall architectural geometry. The bridge is fully accessible with curved approach ramps, which were required to provide the minimum vertical clearance over the roadway. Both access ramps have an S-shaped geometry to minimize impact on an existing neighborhood park near the crossing, which was renovated as part of the project. An innovative protective fencing system infills the arch seamlessly. The East 54th Street Bridge is the first tied-arch bridge over FDR Drive in New York City and an attractive waterfront marker.

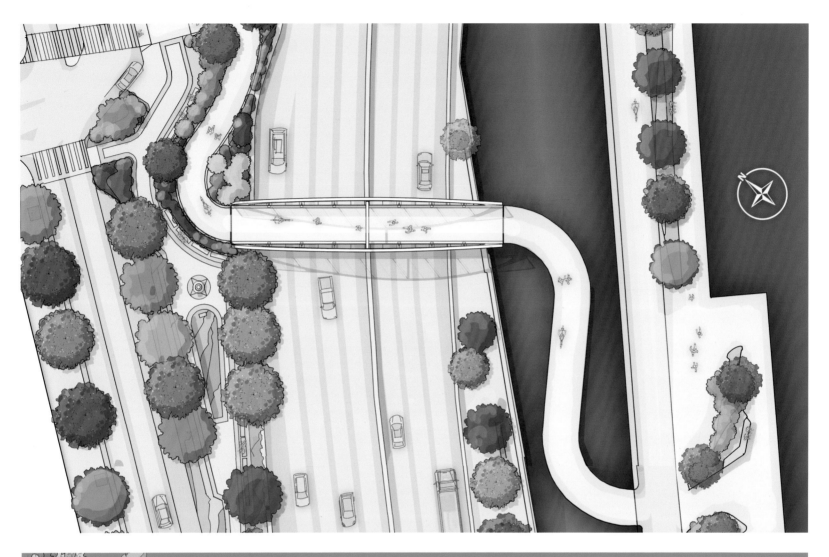

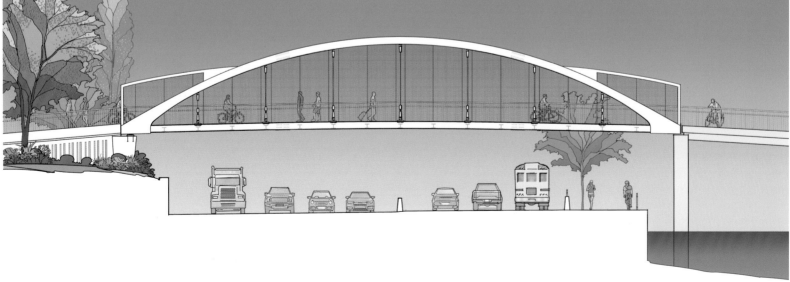

Bridge architectural features and detailing:

- Mid-century inspired architectural features to visually relate to the context.

- Light-gray structural system complements the marine environment.

- Use of curved access ramps to enhance an adjacent park and provide full accessibility.

- Concrete retaining walls enhanced with vertical rustication and angled geometry relating to the main arch.

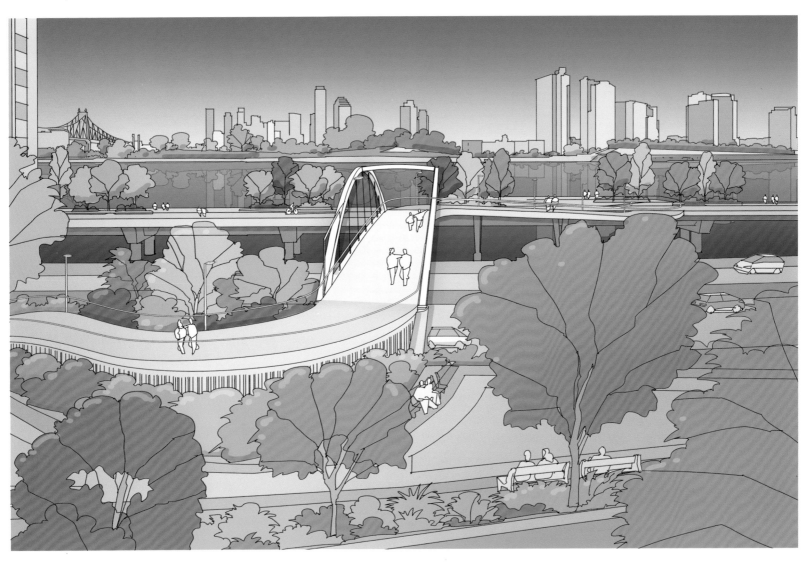

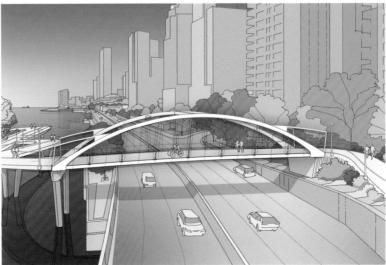

WOODROW WILSON
bridge

WOODROW WILSON
bridge

Location
Washington, DC

Completion date
2008

Width
249' (75.8 m)

Total length
6,000' (1,828.8 m)

Main span length
270' (82.2 m)

Vertical clearance
70' (21.3 m)

Horizontal clearance
175' (53.3 m)

The Woodrow Wilson Memorial Bridge spans the Potomac River between Maryland and Virginia, near Washington, DC. The bridge is named after Woodrow Wilson, the twenty-eighth president of the United States, who was born in Virginia. The bridge carries Interstates 95 and 495 as part of the Capital Beltway, with twelve lanes of traffic on a pair of twin structures. A sidewalk for pedestrians and bicyclists also crosses the entire bridge and connects to parks and trails on both sides of the river.

The bridge design concept was selected via a national competition in 1998. An important design goal in the competition brief was that the new bridge should be visually compatible with existing historic arch bridges in the area. To meet this goal, a series of curved V-shaped piers with connecting tension ties was proposed, which allowed the bridge foundations to not be subjected to horizontal arch thrust loads. This feature reduced its foundation cost while creating an appearance that complements nearby historic arch bridges. The bridge incorporates eighteen sets of curved V-shaped support piers that resemble seagulls in flight. This gives the bridge a sense of lightness and a distinct profile visible from long-range viewpoints. The bridge includes a movable bascule section to allow tall vessel navigation. This portion of the bridge is visually integrated into the overall facade, giving it a seamless appearance. Adjacent riverfront parks and trails allow the areas under the bridge to be used for recreational purposes. The bridge's carefully shaped arched piers are illuminated at night to highlight its main features. The Woodrow Wilson Memorial Bridge is considered part of Washington, DC's national monuments core, and has become a recognizable structure along the riverfront.

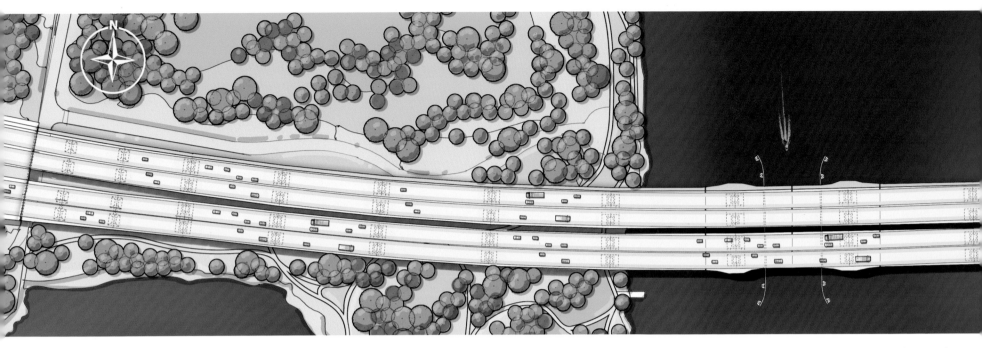

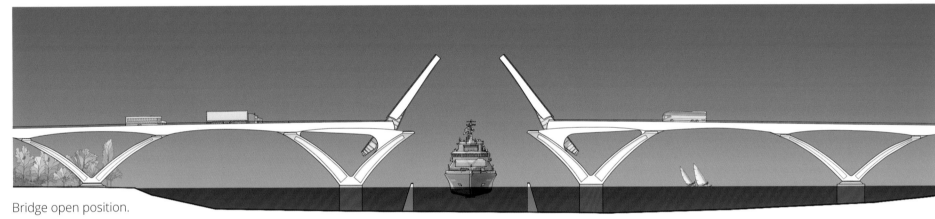

Bridge open position.

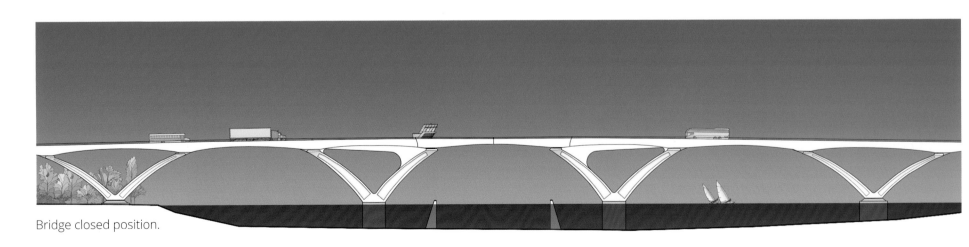

Bridge closed position.

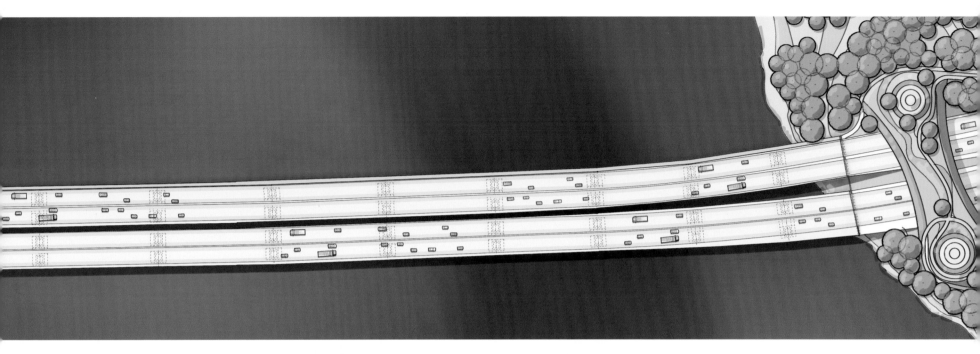

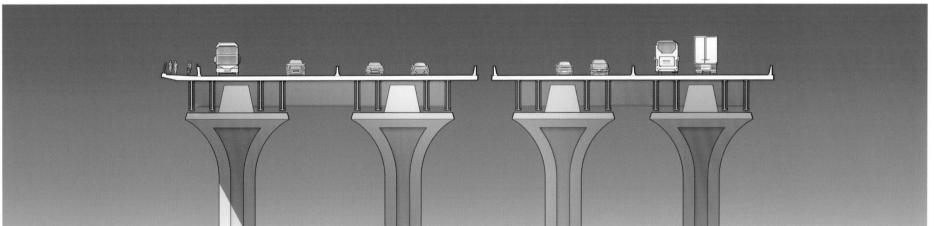

Bridge architectural features and detailing:

- Bascule movable span, shaped to be visually integrated with the rest of the bridge's overall facade.

- Curved V-shaped piers visually compatible with nearby historic arch bridges.

- Arched railings and enhanced movable bridge control tower with curved features.

- Distinct light-gray color used on the entire bridge.

- Aesthetic lighting enhances the bridge at night with changing color capabilities.

COLUMBUS AIRPORT
bridges

COLUMBUS AIRPORT
bridges

Location
Columbus, OH

Completion date
2008

Width of taxiway bridge
217' (66.1 m)

Width of service bridge 1
74' (22.5 m)

Width of service bridge 2
29.5' (8.9 m)

Length of each bridge
191' (58.2 m)

Vertical clearance
20' (6 m)

The John Glenn Columbus International Airport in Columbus, Ohio, needed a crossover taxiway for aircraft traveling the relatively short distance from the terminal building to the outer runways. The taxiway bridge was designed to carry a Boeing 747-400 aircraft weighing 894,900 pounds. The bridge spans the primary entrance roadway to the airport terminal, which includes seven vehicle lanes and two light-rail lines. The exceptionally wide bridge consists of a post-tensioned cast-in-place concrete structural system with integral, inclined abutments. The bridge span has a variable depth profile in addition to reverse tapers on the abutments and enhanced architectural concrete parapets.

Two additional adjacent service bridges were built at the same time to improve connectivity at the airport between service zones and maintenance facilities. The trio of bridges relate visually and include the same structural system and architectural detailing for overall consistency. When approaching the airport, the family of bridges creates a harmonious, welcoming, and recognizable ensemble. The bridges were built supported by the ground soil. This was feasible as the initial ground-level roadway was approximately at the same level required for the underside of the planned bridges. This construction technique reduced the cost of the bridges and expedited construction. A system of attractive blue linear lights was integrated into the bridges' undersides and inclined abutments. The Columbus Airport Bridges create a gateway into the regional airport with understated architecture, a streamlined appearance, and overall striking slenderness.

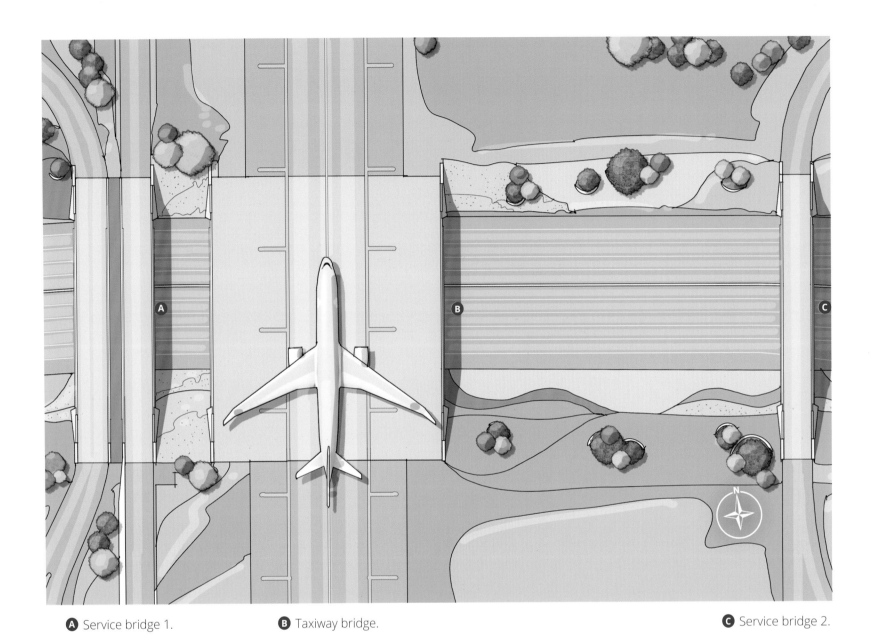

Ⓐ Service bridge 1.   Ⓑ Taxiway bridge.   Ⓒ Service bridge 2.

Bridge architectural features and detailing:

- Family of visually consistent concrete bridges with a streamlined appearance.

- Inclined and tapered abutments with horizontal rustications.

- Distinct light-gray color for visual consistency.

- Aesthetic lighting enhances the bridge at night with blue linear lights integrated into the abutments and bridge deck underside.

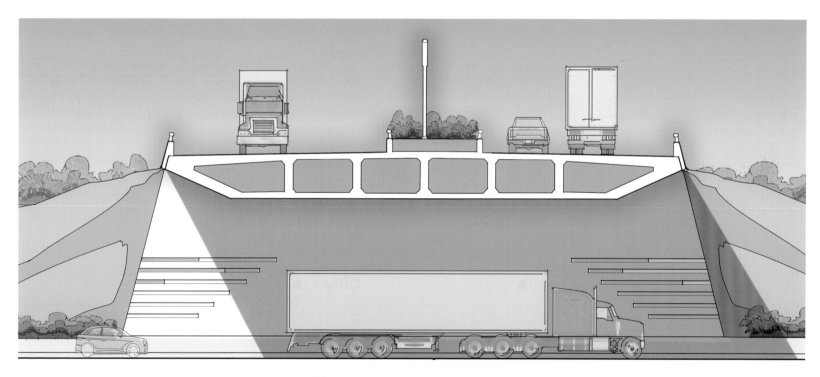

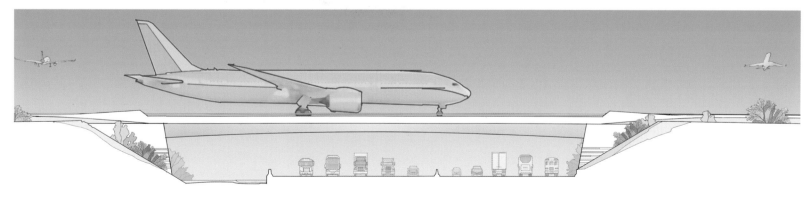

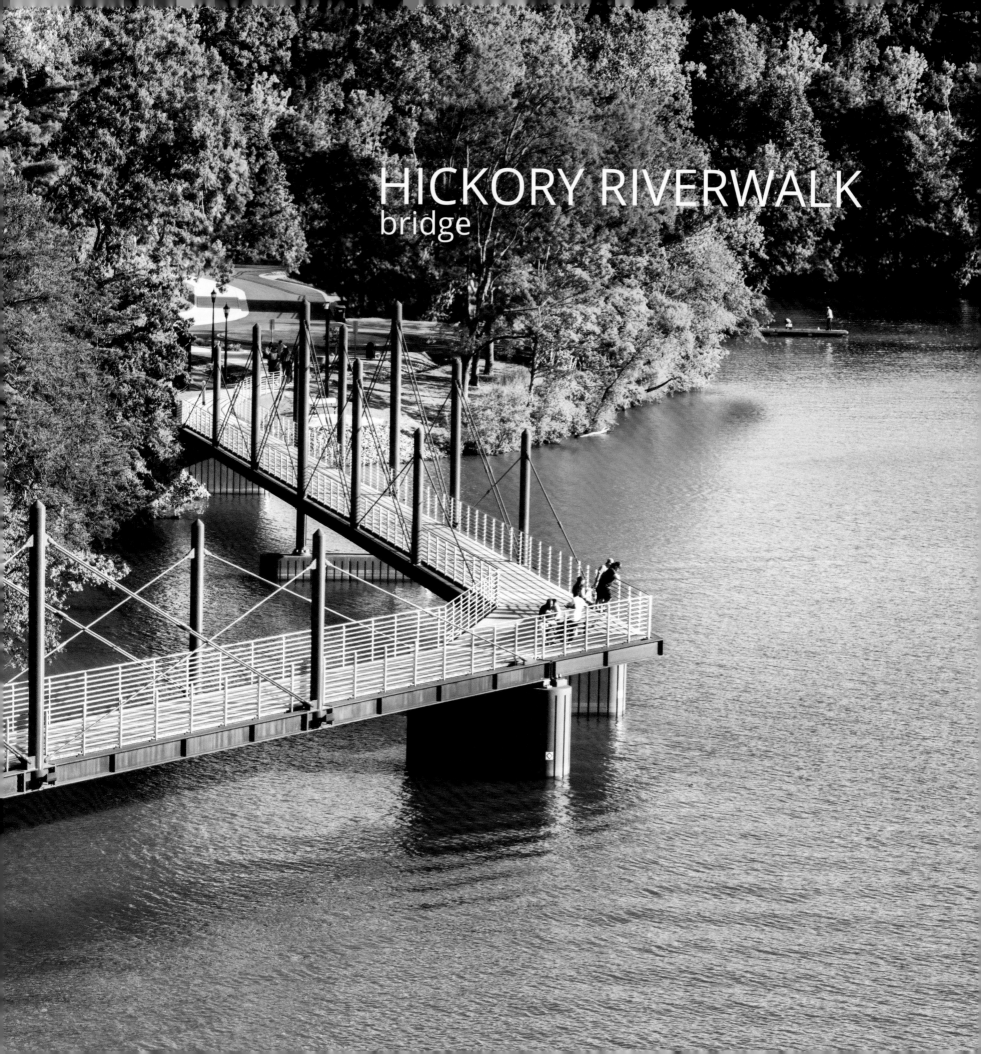

HICKORY RIVERWALK
bridge

HICKORY RIVERWALK
bridge

Location
Hickory, NC

Completion date
2024

Width
10' (3 m)

Total length
520' (158.4 m)

Height of tallest tower
31' (9.4 m)

This riverwalk pedestrian bridge allows the residents of Hickory, North Carolina, to engage with its waterfront and lake. The bridge connects an improved local park to other previously inaccessible zones next to the lake as part of a new trail system. The bridge has become a community asset, generating economic development in the area due to its high visibility and distinct architectural profile.

The pedestrian bridge consists of an inverted Fink truss system that is unique in the region and the longest of its kind in the United States. This type of bridge is based on a truss system patented by Albert Fink in 1854. The Fink truss structural system was used for several decades on bridges owned by American railroad companies across the country in the mid to late 1900s but it has not been used frequently since then. The bridge structural system includes multiple diagonal members projecting up to top posts at a variety of angles, giving the bridge a sawtooth appearance when viewed from the side. The truss also lacks a top chord, as only a bottom chord is needed due to its configuration. The shape of the bridge is reminiscent of maritime docks with its multiple towers of varying heights. The dark brown-reddish color of the steel structure blends well with the existing natural environment and helps integrate the structure with its context. A seating area is located at a central overlook facing the lake. Subtle aesthetic lighting and a wooden walking deck enhances the experience for pedestrians. The Hickory Riverwalk Bridge has become an exciting destination for residents to enjoy views of the lake.

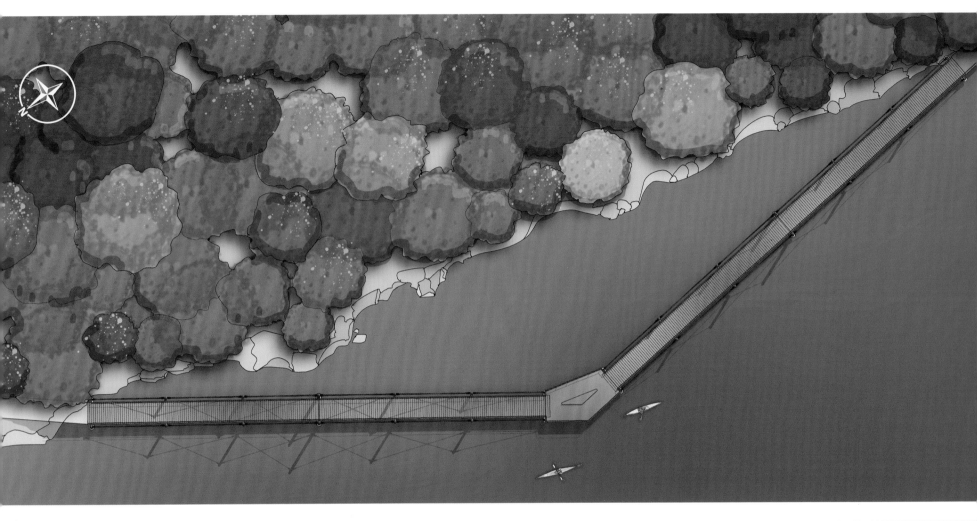

Bridge architectural features and detailing:

- Longest inverted Fink truss bridge in the United States.

- Use of weathering steel and compatible earth-tone coatings to visually integrate into the natural context.

- Triangular overlook and seating area facing the lake.

- Use of sustainable wood decking.

- Shape of towers inspired by the classic design of docks.

- Aesthetic lighting enhances the bridge at night.

Invitation to Hickory Riverwalk Bridge opening event on April 4, 2024.

LIBERTY
bridge

LIBERTY
bridge

Location
Greenville, SC

Completion date
2004

Width
12' (3.6 m)

Total length
380' (115.8 m)

Tower heights
90' (27.4 m)

**Clearance
above the river**
50' (15.2 m)

The Liberty Bridge is a suspension pedestrian bridge overlooking a group of waterfalls and gardens at Falls Park in Greenville, South Carolina. The bridge replaced a six-lane highway overpass that was demolished during the redevelopment of the area's public spaces and parks. A highway bridge had covered the waterfalls for many decades, obscuring their beauty and historical importance. The new bridge was moved away from the falls to help create a visual connection between them. Its main span over the Reedy River arcs away from the falls and provides visitors with an aerial platform and overlook from which to view the cascading water and adjacent parkland.

The bridge gently slopes into the ravine, supported by twin inclined steel towers, a curved truss, and a single suspension cable with inclined cable hangers. The bridge carefully merges into the park's circulation system and blends well with the existing topography. The bridge appears to float over the landscape seamlessly, integrating with the surrounding tree canopy. The twin inclined towers and suspension cable are visible from several vantage points around the city, drawing visitors to the public park, falls, and river during the day and night. As a local landmark, it has become a popular place for couples to get engaged. The suspension bridge is named after the Liberty Corporation, which provided substantial funding for its completion and future maintenance. The bridge became a catalyst for the redevelopment of the downtown and expansion of its main commercial street in the long-dormant West End neighborhood. Since its completion, the Liberty Bridge has brought substantial development and prosperity to the area, becoming a contemporary symbol of the city and the region.

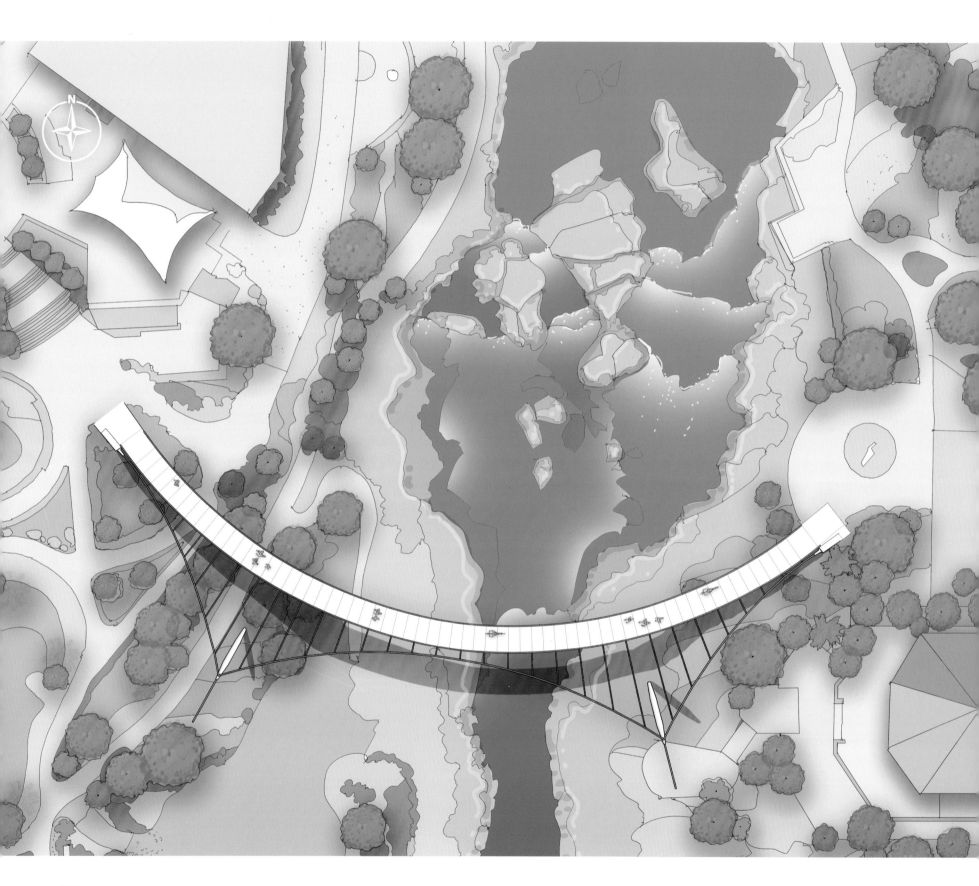

Innovations and technological advancements at the time of completion:

- First curved suspension bridge with a single suspension cable in the United States.

- Use of ring cables that provide compression in the horizontal plane to achieve a balanced structure in conjunction with the hanger and suspension cables above the deck.

- Use of a curved steel truss underneath the deck that addresses torsional loads as the bridge is supported by cables on only one side.

Bridge architectural features and detailing:

- Slender pedestrian railings with razor-thin cables for unobstructed views and overall transparency.

- Well-proportioned, curved structural truss superstructure with angled and curved elements.

- Tapered, inclined steel towers with bottom sphere assemblies to allow for bridge flexibility and thermal movements.

- Integrated railing lighting and overall aesthetic lighting that highlights the towers and cables.

- Carefully detailed bridge abutments, cable anchors, and connections to enhance elegance.

- Light-colored structural elements to highlight the bridge within the park's context.

THROOP STREET
bridge

THROOP STREET
bridge

Location
Chicago, IL

**Estimated
completion date**
2027

Width
80′ (24.3 m)

Total length
315′ (96 m)

Height of arch
111′ (33.8 m)

Navigation clearance
18.5′ (5.6 m)

The Throop Street Bridge in Chicago, Illinois, will connect two portions of the Lincoln Yards development across the North Branch of the Chicago River. Lincoln Yards is a new fifty-three-acre mixed-use project that will be developed in a former industrial zone along the river. The overall project will add twenty-one acres of open and publicly accessible space.

The bridge will accommodate vehicles, bicyclists, and pedestrians. Pedestrian sidewalks have been placed at the outer edges of the bridge with views of the water. Bicyclists have been separated from vehicular traffic with a safety barrier. A cycle track on the bridge will accommodate an extension of a regional bike trail that replaced a former rail line running east-west on the northwest side of Chicago. The arch bridge is inspired by historic bridges over the Chicago River and includes a lattice truss system, rusticated abutments, and V-shaped railings. Crossing the river without any piers in the water will allow the restoration of the riverbanks. There will also be wide walkways under the bridge connecting to parks to be built as part of the development. The bridge underside will include two zones in which the deck will be open to the sky, allowing sunlight to reach the river to reduce its visual mass. At night the illuminated bridge will become a beacon for the community and a symbol of the development that will transform the former industrial area. Its prominence and distinct light-blue color will mark the river crossing from a distance, making the Throop Street Bridge a signature component of the multibillion-dollar Lincoln Yards project.

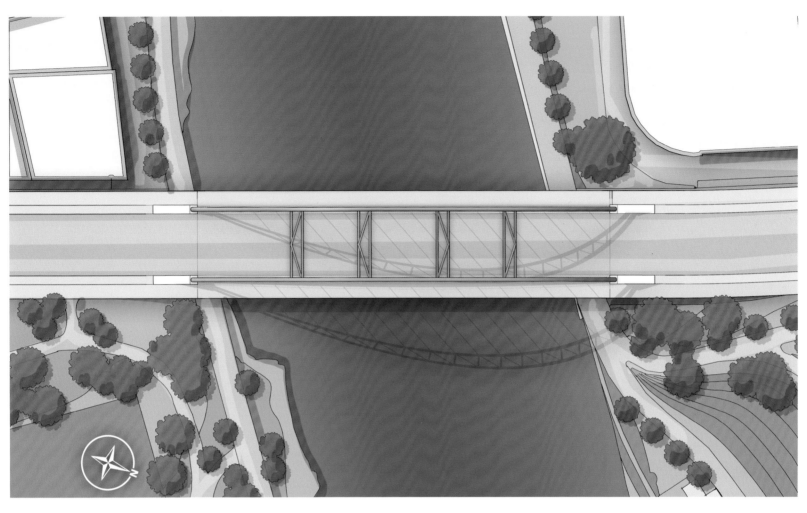

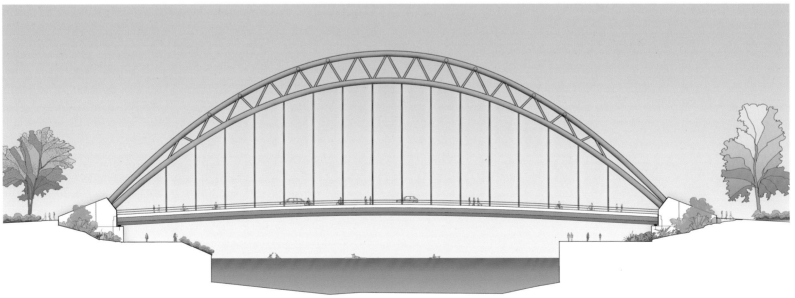

Bridge architectural features and detailing:

- Arch inspired by traditional steel bridges across the Chicago River.

- Well-proportioned structural truss superstructure with angled elements.

- Integrated aesthetic lighting that highlights the arch.

- Carefully detailed bridge abutments.

- Light-blue structural elements and matching railings that enhance the bridge from a distance.

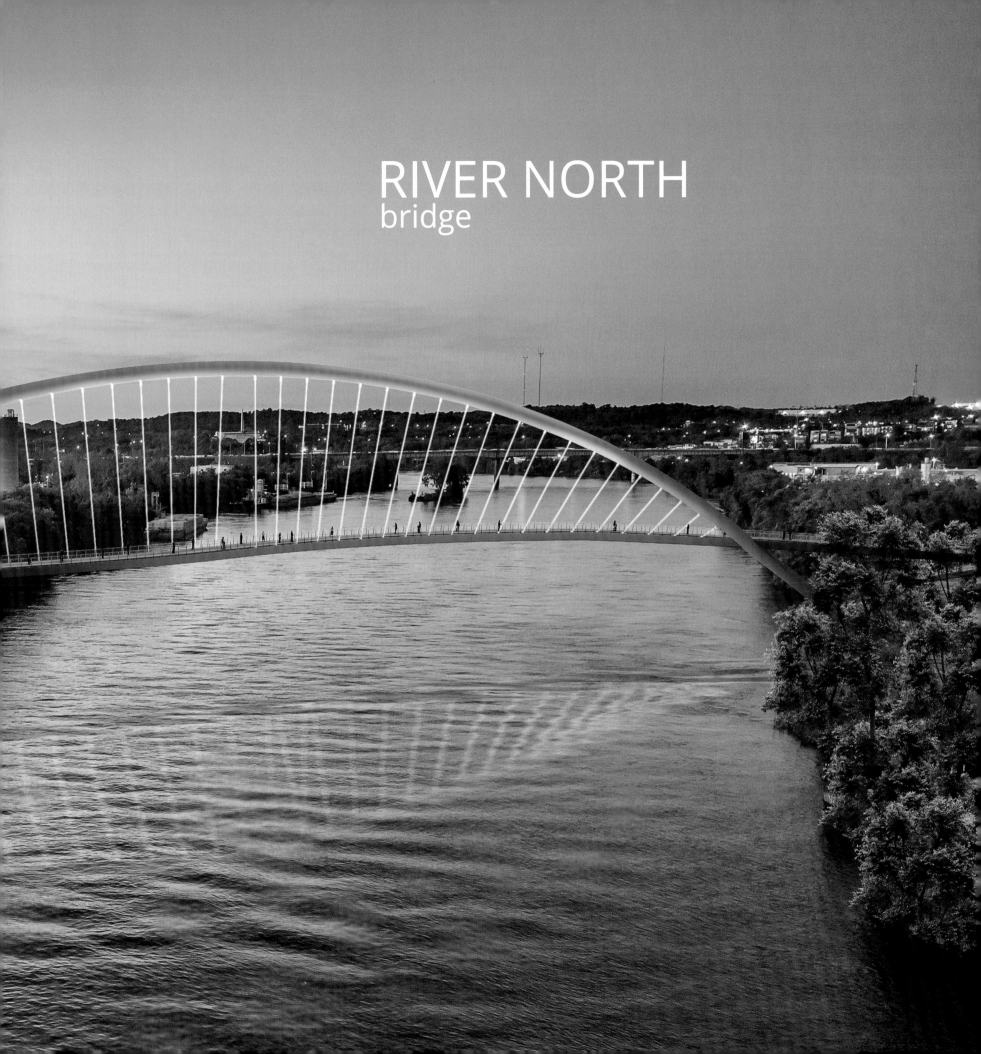

RIVER NORTH
bridge

RIVER NORTH
bridge

Location
Nashville, TN

**Estimated
completion date**
2027

Width
16' (4.8 m)

Total length
1,100' (335.2 m)

Main span length
750' (228.6 m)

Height of arch
193' (58.8 m)

Navigation clearance
70' (21.3 m)

The bridge will connect a new campus for Oracle Corporation to Germantown, a residential neighborhood across the Cumberland River near downtown Nashville, Tennessee. The River North Bridge will help transform the East Bank of the Cumberland River by creating a signature structure visible from the new campus buildings and adjacent neighborhoods.

The proposed bridge will cross the river on a diagonal, creating a fluid and exciting journey across the river with a variety of changing views of the riverfront and cityscape. An arch with an S-shaped configuration complements and balances an S-shaped bridge alignment, creating an infinity symbol that appears as a sideways number eight when seen from above. The highest point of the arch will be located at the center of the river. The bridge will have a clear span with no supports in the water, which will improve constructability and allow for ease of navigation. The arch will frame the entire width of the river and mark the crossing from a distance. The bridge includes a torsional-resistant structural box as cables connected to the three-dimensional arch will only be attached from one side of the bridge deck. A long curvilinear ramp will be required to reach grade for accessibility on the east side. The ramp will use similar architectural details as the bridge and will be visually transparent in order to not obstruct the views of the Cumberland River and a proposed riverfront greenway. Subtle lighting, elegant cable anchors, and slender railings enhance the bridge's appearance at close range. With its daring S-shaped arch, the River North Bridge will be alluring, memorable, innovative, and structurally efficient.

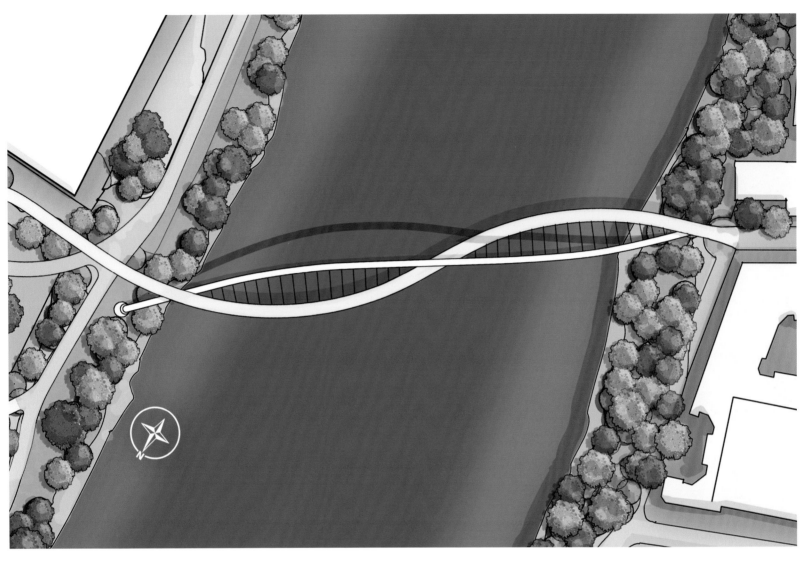

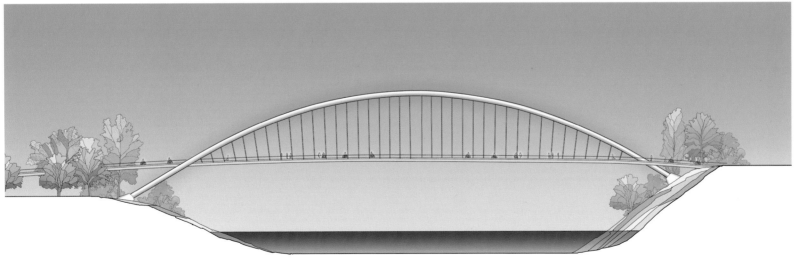

Innovations and technological advancements:

- First curved 3D-arch bridge with an S-shaped layout in the United States.

- Use of a triangular steel box to provide torsion resistance for the S-shaped alignment.

- Longest span arch pedestrian bridge over the Cumberland River.

- Integrated curved access ramp with tall, slender piers.

I-74 MISSISSIPPI RIVER
bridge

Location
Quad Cities, IA and IL

Completion date
2022

Total width of twin bridges
241' (73.4 m)

Width of walkway
14' (4.2 m)

Main span length
800' (243.8 m)

Height of arches
250' (76.2 m)

Navigation clearance
60' (18.2 m)

The I-74 Mississippi River Bridge is located near the geographic center of the Iowa-Illinois Quad Cities region, connecting Bettendorf, Iowa, and Moline, Illinois. New twin arches replaced a pair of suspension bridges that were built in 1935 and 1961 as part of a seven-mile expanse comprising the main river crossing, ramp approaches, and associated intersections. It was important to replace the historic bridges as they were functionally obsolete and not compliant with current code requirements.

The new bridge is taller than the original ones and more visible from a distance due to its distinct twin basket-handle arches and a wider navigation channel. The slender arches are the longest basket-handle true arches erected in place in the United States. Each of the arches includes four lanes of traffic plus safety shoulders. One of the arches has a wide walkway on the outer side of the main arch that connects pedestrians and bicyclists to trail systems on both sides of the river. The walkway also includes a dramatic overlook area with seating for resting and viewing the waterfront. The twin arches are slender and have a distinct profile, appearing as an ensemble from a distance. A minimum amount of bracing between the arch ribs was included for a lighter, airy appearance. The arches visually terminate at river level to enhance visual appeal. The overall bridge detailing, including the roadway lampposts and pedestrian railings, relates to the curved configuration of the arch structures. The I-74 Mississippi River Bridge has become the region's defining landmark and a source of community pride and inspiration for future generations.

On August 24, 2023, the United States Postal Service issued a stamp celebrating the design and engineering of the I-74 Mississippi River Bridge.

Bridge architectural features and detailing:

- Twin steel basket-handle arches.

- Minimal structural bracing between arches for visual clarity.

- Protected pedestrian and bicycle walkway along the entire bridge.

- Aesthetic lighting highlights the arches and approach spans.

- Elegantly and carefully detailed bridge arch abutments, cable anchors, and connections.

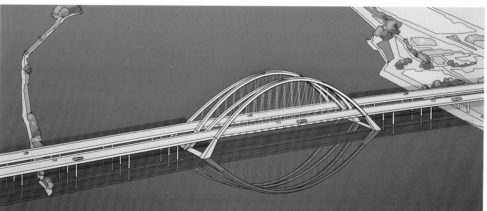

HUNTSVILLE
bridge

HUNTSVILLE
bridge

Location
Huntsville, AL

Estimated completion date
2027

Width
12' (3.6 m)

Total length
1,390' (423.6 m)

Height of tower
118' (35.9 m)

Vertical clearance
17.5' (5.3 m)

The Huntsville Bridge will be an iconic pedestrian and bicycle bridge in downtown Huntsville, Alabama. The curvilinear bridge crosses over Memorial Parkway, a highway that traverses the city north-south. The elevated highway has acted as a physical and visual barrier between residential neighborhoods west of downtown and the jobs-centric office district on the east side. The new bridge connects the Von Braun Center and business district to the Lowe Mill ARTS & Entertainment complex, a privately owned regional arts facility that hosts multiple art studios and related local businesses in a historic textile mill.

The bridge consists of a series of three inclined steel towers with catenary suspension cables that are in visual and structural balance with the overall curvature of the bridge alignment. The three groups of cables are attached to only one side of the bridge deck creating a distinct geometry. The suspension bridge has a curvilinear footprint to avoid obstacles along the crossing, including several highways and local streets. Canals in the vicinity of the bridge will be enhanced and landscaped as part of an overall plan. The design includes visually transparent railings, integrated aesthetic lighting, and curved fencing over the roadways. The Huntsville Bridge will be easily visible from a distance, marking and celebrating Huntsville's downtown and the city's cultural and business core.

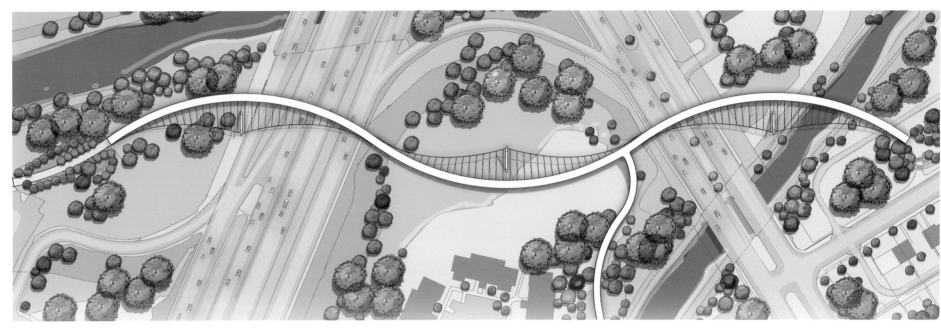

Bridge architectural features and detailing:

- Elegant, continuous box steel superstructure with tapered cross section.

- Aesthetic lighting enhances the bridge at night by highlighting three main towers.

- Elimination of back span anchor cables creates visual balance between the curvilinear bridge alignment and inclined towers.

- Curved fencing details to enhance bridge crossings over the highways.

GRIFFIN
bridge

GRIFFIN
bridge

Location
Des Moines, IA

Completion date
2006

Width
12' (3.6 m)

Main span length
230' (70.1 m)

Height of arch
62' (18.9 m)

Vertical clearance
17.5' (5.3 m)

The Edna M. Griffin Memorial Bridge is located over the busiest highway in Iowa, near downtown Des Moines. The blue basket-handle arch bridge is one of a family of three arch bridges that were built at separate locations during the reconstruction of a section of Interstate 235. The bridges mark city neighborhoods, schools, and the Greater Des Moines Botanical Garden along the expressway. Given how fast highway drivers cross this portion of the city, the idea was to create a family of distinctive bridges, one after the other, that would define a sense of place along the fourteen-mile route. This project illustrates how bridges can help give identity, enhance the environment, and connect areas of cities previously separated by highways.

The basket-handle arches consist of tapered steel boxes that connect at the center, with minimum bracing between them for structural clarity and simplicity. This structural arch system was chosen because the steel arches could be built off-site. This allowed them to be placed over the highway at night in a short period of time, minimizing traffic impacts during construction. A curved screen fencing system, in a contrasting light color, was used to enhance safety as well as appearance. All three bridges are aesthetically illuminated and act as community beacons. The bridges have helped improve growth and renewal across the city and become markers along the highway corridor.

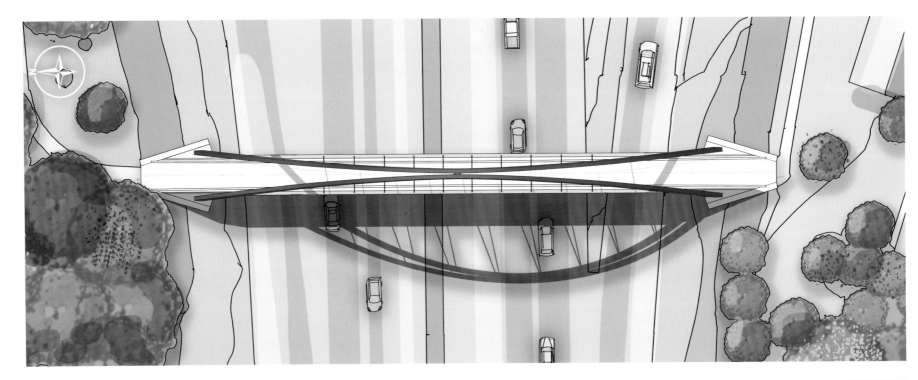

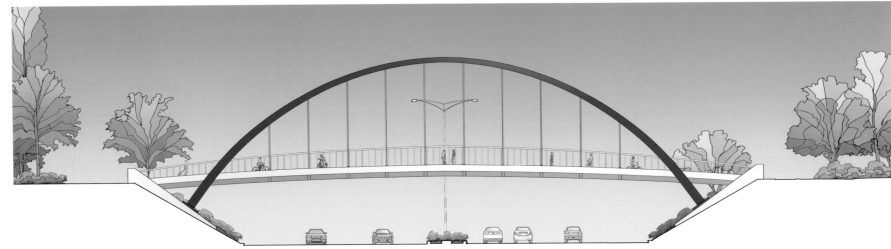

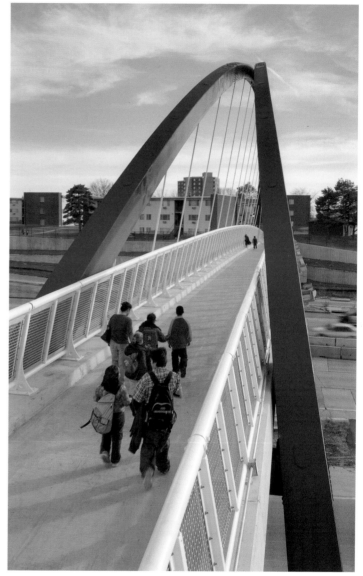

Bridge architectural features and detailing:

- Arched railings and fences integrated into the overall arch form.

- Distinctive blue color visible from long distances to mark the crossings over the highway.

- Aesthetic lighting enhances the bridge at night.

- Minimum use of bracing between arch ribs for visual clarity and structural simplicity.

CARVER
bridge

CARVER
bridge

Location
Des Moines, IA

Completion date
2005

Width
105' (32 m)

Total length
279' (85 m)

Height of arch
59' (17.9 m)

The George Washington Carver Bridge in Des Moines, Iowa, crosses over the Raccoon River. The bridge is an integral part of the Martin Luther King Jr. Parkway that traverses the city, mainly at grade. The parkway replaced a proposed elevated highway that would have blocked views of the city and river and severely impacted the central business district. The bridge has three roadway lanes in one direction and two roadway lanes and one bikeway/pedestrian lane in the other. The structure is named after George Washington Carver, a prominent Black agricultural scientist who was renowned for inventing crop rotation and helping farmers improve their output.

The main river crossing features two blue free-standing arches that span the entire river without piers in the water. The slender arches act as a gateway when approaching the city from the airport, framing views of the skyline and Iowa's tallest downtown skyscraper. The distinct color of the arches complements the site's abundant greenery along the river. Clearly marking the river crossing, the arches appear to float over the water from a distance. The arch ties were built with post-tensioned, cast-in-place, reinforced concrete with a smooth bridge underside. The arches were furnished in three pieces and welded together in the field with seamless connections, improving the visual quality and detailing of the bridge. Pedestrian railings, roadway barriers, and lighting fixtures match the color of the main arches to reinforce its overall visual appeal. The George Washington Carver Bridge has become a Des Moines icon. In 2008 the city incorporated a silhouette of the arch in its official logo to celebrate the transformation of downtown that followed the completion of the bridge and parkway.

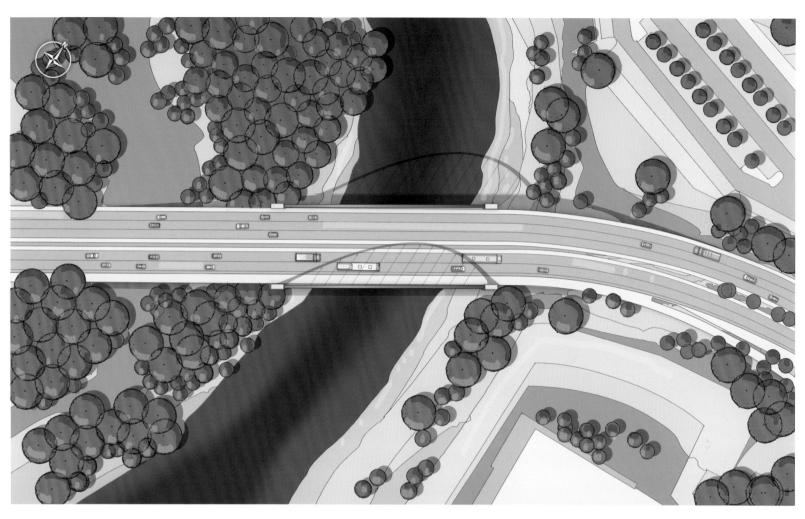

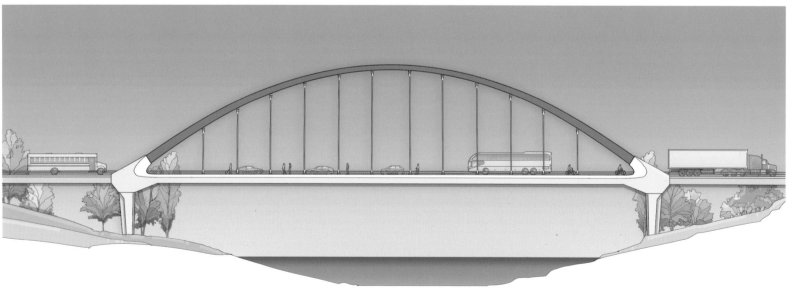

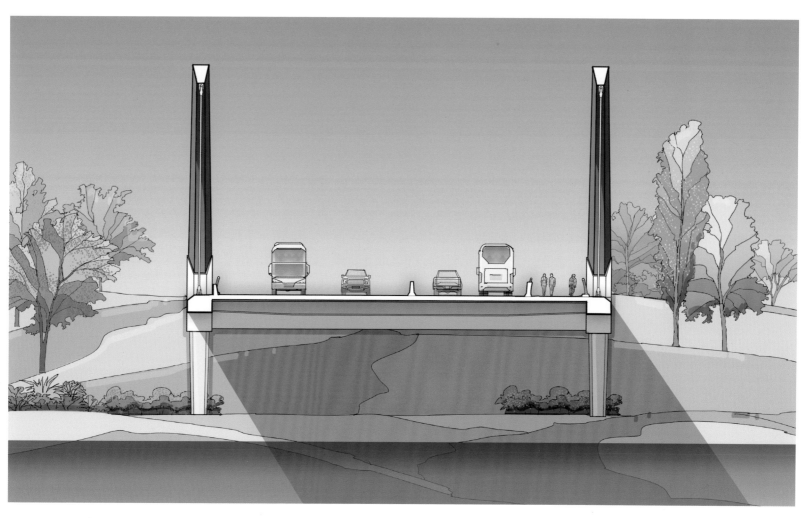

Bridge architectural features and detailing:

- Pedestrian railings integrated into the overall arch form.

- Distinct blue color visible from long distances to mark the crossing over the river.

- Absence of bracing between arch ribs for visual clarity, structural simplicity, and to frame city views.

CITY OF **DES MOINES**

City of Des Moines logo inspired by the George Washington Carver Bridge.

BRUCE VENTO
bridge

BRUCE VENTO
bridge

Location
Saint Paul, MN

Estimated completion date
2027

Width
12' (3.6 m)

Total length
1,475' (449.5 m)

Main span length
495' (150.8 m)

Height of tower
110' (33.5 m)

The Bruce Vento Bridge will be located near downtown Saint Paul, Minnesota, next to the Mississippi River. The pedestrian bridge will connect two major regional trails in the vicinity of the Bruce Vento Nature Sanctuary. It will cross over a local roadway connector (Warner Road) and multiple railroad lines that have separated the city's East Side community from the river for decades.

The bridge will preserve views of the nearby bluffs, minimize impacts on the adjacent nature sanctuary, and improve views of the sacred site of the Dakota tribe, known as Carver's Cave. Due to the presence of a small regional airport along the river, the bridge has a not-to-exceed height limitation. It also must allow for the future expansion of high-speed passenger rail lines, which require increased vertical clearances. A single inclined steel tower over the rail lines provides a visual anchor along the river. The location of the tower was carefully considered in order to not obstruct views of the bluffs and, at the same time, not impact railroad operations. The curved geometry of both access ramps and the main crossing allows for a variety of changing views of the mountains, nature preserve, and river. A convenient curved staircase next to the river provides faster access when the fully accessible ramps are not needed. The distinct bridge cable arrangement is suspended only from the interior side of the curvature of the deck, creating a visually appealing structure that enhances the area. The Bruce Vento Bridge will be illuminated at night for an enriched experience and to create a symbolic gateway into the city.

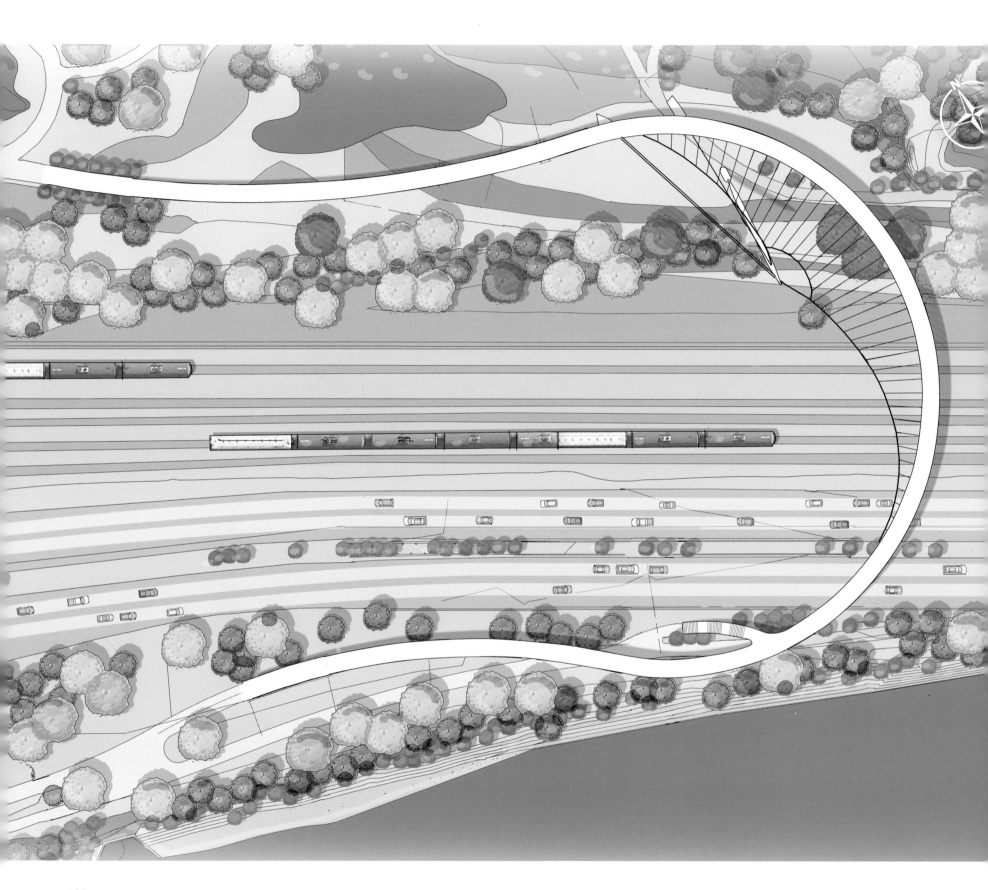

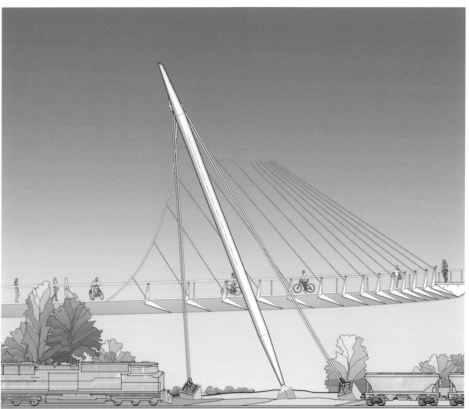

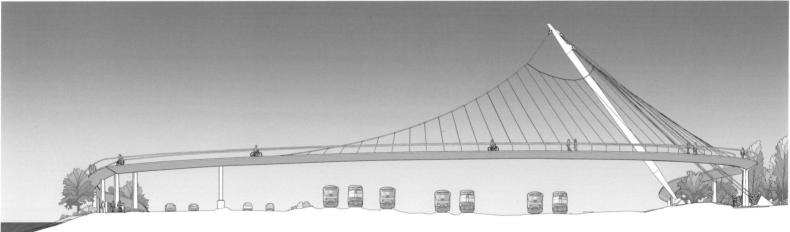

Bridge architectural features and detailing:

- A single inclined and tapered tower marks the bridge from a distance.

- Single suspension bridge with thin cable hangers that provide transparency and enhance the riverfront.

- Slender railings and fence detailing for open views of the river, nature sanctuary, and historic sites.

- Visually consistent box superstructure along the entire bridge and approach ramps.

- Aesthetic illumination of the tower and cables.

- Curvilinear bridge layout in harmony with setting and nature preserve.

COMO PARK
bridge

COMO PARK
bridge

Location
Saint Paul, MN

Completion date
2014

Width
15' (4.5 m)

Total length
88' (26.8 m)

Main span length
50' (15.2 m)

The Como Park Bridge, originally built in 1904, is in historic Como Park in Saint Paul, Minnesota. It used to cross over a streetcar service line, which was discontinued in 1954. For decades the pedestrian bridge was rarely used and fell into disrepair as it had lost its functionality and was no longer needed as an above-grade separation. Before its restoration the bridge was crumbling, deserted, and severely distressed.

The three-span, open-spandrel, reinforced concrete arch is an early example of the Melan arch structural system. This type of arch structure uses arched steel beams set in concrete. A curved metal lattice beam supports the concrete while it hardens. After the bridge is completed, the concrete arch carries the load. The innovative structural system was named after Austrian engineer Josef Melan, who conceived it. Melan arch bridges could be constructed quickly and were stronger than many other designs, leading to their widespread use. The Como Park Bridge was rehabilitated to meet current standards. The main arch was repaired and strengthened. The restored bridge includes new balustrades that replicate the originals with slight modifications to comply with current structural guidelines. New paths were built under and over the bridge to reconnect it to the park trail system. Interpretive signage was added to inform the public about the bridge's significance in terms of its engineering, original purpose, and setting. The public uses the restored bridge as an overlook in the park. Aesthetic lighting was installed under the main arch to highlight it at night. The fully restored Como Park Bridge with its new function in the park has become an asset for the city.

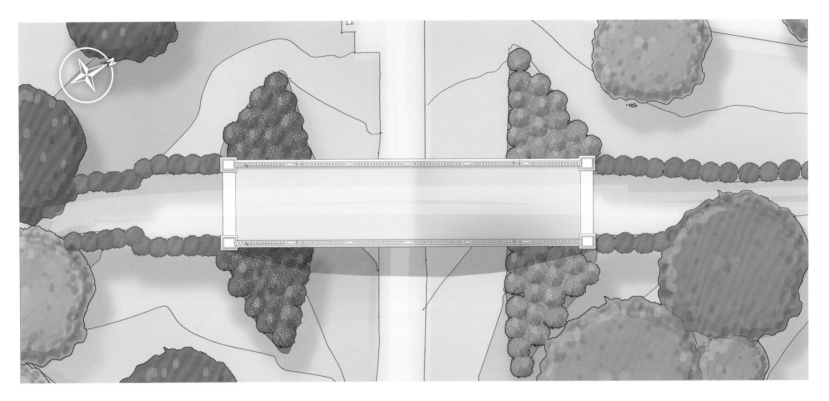

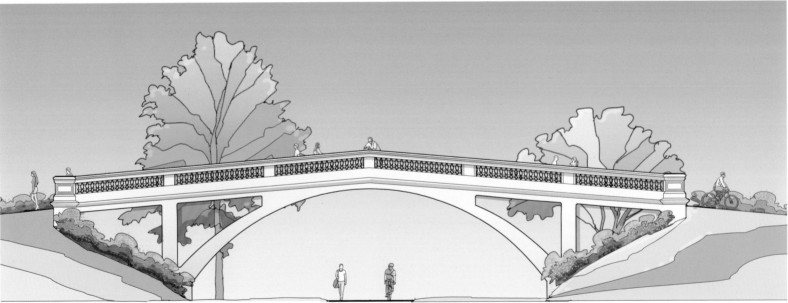

Historical character-defining bridge elements that were restored, replicated, or enhanced:

- The Melan concrete arch was repaired and strengthened. The concrete surfaces treated to achieve a more consistent appearance.

- New sets of concrete balustrades closely match the original design while complying with current codes.

- New lighting was added to highlight the shape of the concrete arch at night.

- New pedestrian paths added over and under the bridge to create a park overlook, giving new functionality to the bridge.

The Como Park Bridge was opened in 1904 and for decades crossed over a streetcar line that traversed the park. After the streetcar was discontinued in 1954, the bridge lost its purpose and fell into disrepair. These photos illustrate the advanced state of deterioration prior to its restoration and rehabilitation.

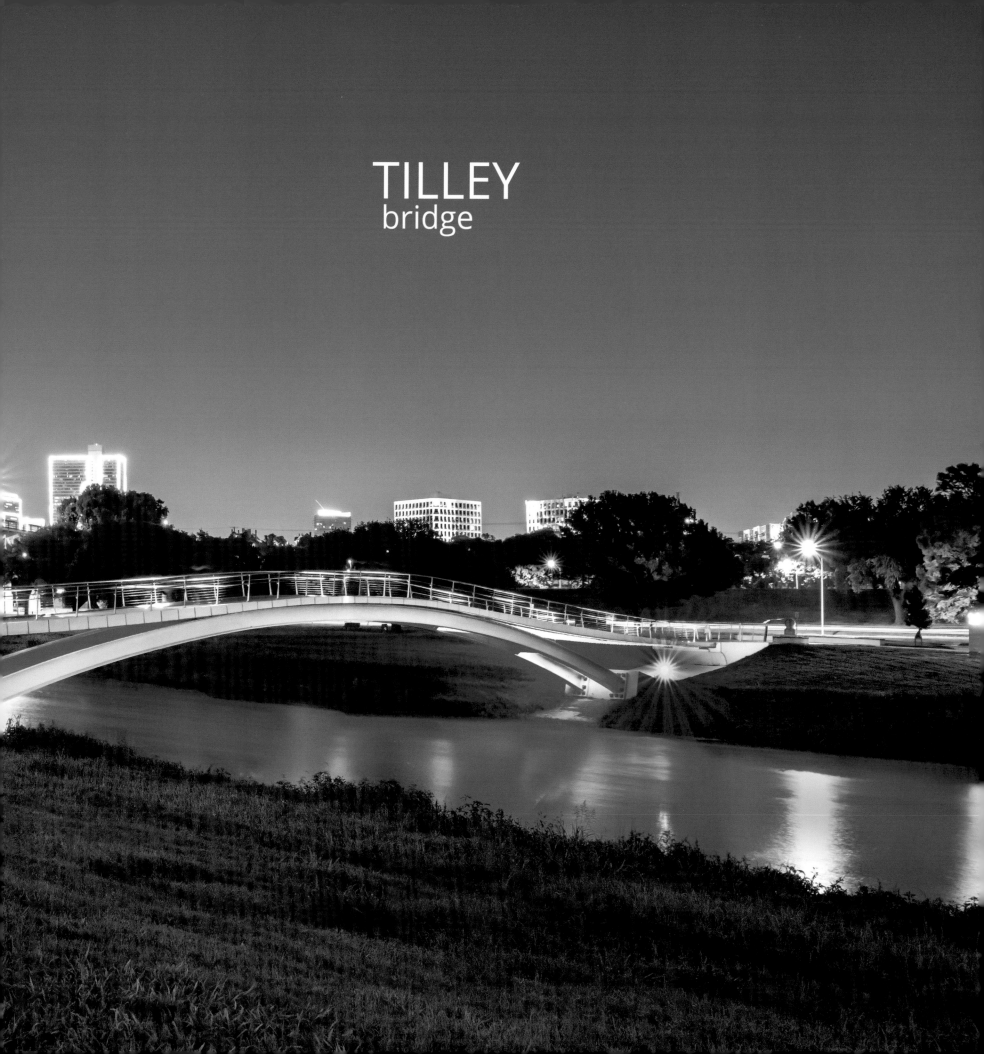

TILLEY
bridge

TILLEY
bridge

Location
Fort Worth, TX

Completion date
2012

Width
12' (3.6 m)

Total length
368' (112.1 m)

Main span length
163' (49.6 m)

The Phyllis J. Tilley Memorial Bridge crosses over the Trinity River and connects Trinity Park to a new trail that terminates in downtown Fort Worth, Texas. The pedestrian bridge has a graceful profile that enhances the serene landscape along the river. The river crossing was named after Phyllis J. Tilley, an advocate for the use of the riverfront and a founding member of Streams & Valleys, Inc., a community organization that contributed financially to help build it.

A steel arch spanning the entire river supports steel stress-ribbon segments and precast concrete planks. The bridge shape complements the adjacent historic Lancaster Avenue vehicular bridge. Pedestrians and bicyclists crossing the bridge experience a smooth, undulating, fully accessible walking surface. The steel arch frames the river, and the steel ribbons sit airily over the rounded arch, appearing to flow toward the bridge abutments. The absence of vertical struts between the arch and the bridge deck reduces the horizontal loads created by periodic river flooding. One of the challenges with stress-ribbon bridges is that they require steep undulating slopes, which makes meeting accessibility requirements difficult. The solution was to design a precast concrete deck panel system, of varying thicknesses, to create a walking surface with a series of short ramps and landings that meet the accessibility slope requirements. This feature is not readily apparent along the bridge facade. At night the bridge is illuminated with a combination of white and blue LED lighting for increased safety and aesthetic appeal. The Tilley Bridge has enhanced the beauty of the Trinity River and its potential for community, recreational, and contemplative pursuits and activities.

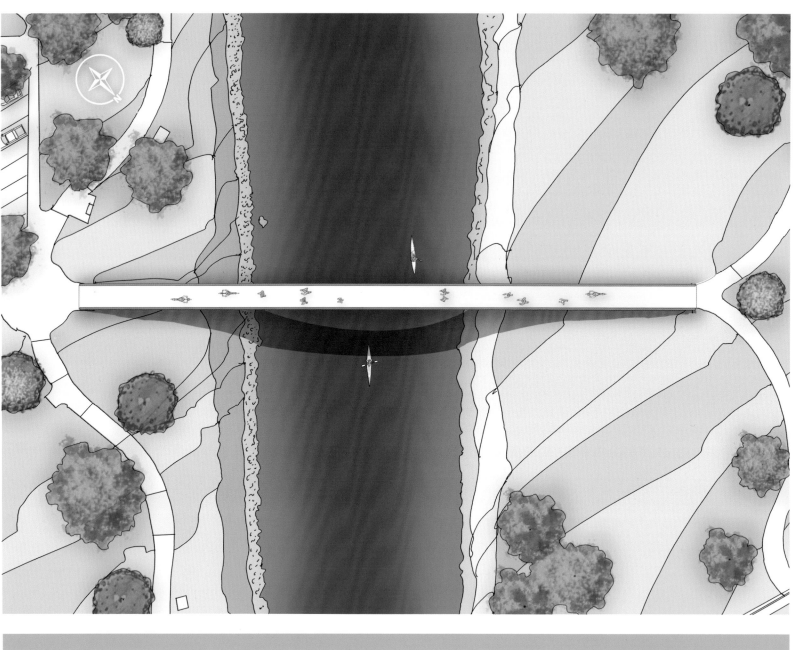
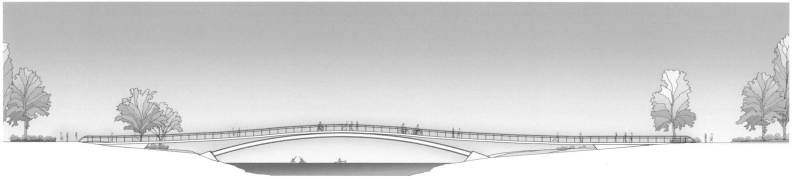

Innovations and technological advancements at the time of completion:

- First arch/stress ribbon combination bridge built in North America.

- Use of a precast panel system directly attached to stainless steel stress ribbons to create a walking surface.

- Balanced and efficient bridge foundations: the minimal sag profile of the stressed ribbons creates major tension loads that are balanced against the thrust loading in the opposite direction caused by the steel arch.

Bridge architectural features and detailing:

- Well-proportioned, slender steel arch and curved profile over the river.

- Slender pedestrian railings with razor-thin cables for unobstructed views and overall transparency.

- Elegantly and carefully detailed bridge abutments for both arch and stress ribbons, visually connected with an inclined wall.

- Light-colored structural elements that highlight the bridge within the park's context along the riverfront.

PANTHER ISLAND
bridges

PANTHER ISLAND
bridges

Location
Fort Worth, TX

Completion date
2021

Widths
White Settlement: 80' (24.3 m)
Henderson: 76' (23.1 m)
North Main: 80' (24.3 m)

Total lengths
White Settlement: 949' (289.2 m)
Henderson: 807' (245.9 m)
North Main: 449' (136.8 m)

Main span length
175' (53.3 m)

Railroad clearance
23' (7 m)

The Panther Island Bridges, part of the Trinity River Vision project, are next to downtown Fort Worth, Texas. When the entire project is completed, they will cross over a future bypass river channel. The bridges replaced existing at-grade crossings at White Settlement Road, Henderson Street, and North Main Street, all located between the city's business core and the Stockyards historic district. The bridges are part of a large project that will double the size of the central business district. The former industrial area, where the new bridges are located, will be redeveloped to include a canal system, walking trails, a marina, and a houseboat district.

The bridges provide a pedestrian-friendly experience with wide sidewalks and transparent pedestrian railings. Their distinct V-shaped piers complement the iconic architectural elements of the city's Cultural District and nearby museums. The design is streamlined and modern, yet suggestive of the arches of the nearby historic Lancaster Avenue and Main Street bridges.

The bridges touch the landscape gently by minimizing the number of piers required to achieve longer spans. The bridge superstructure consists of trapezoidal concrete boxes with clean lines and uncluttered bridge undersides. The cast-in-place, post-tensioned box girders include wide cantilevered sidewalks that help create a lighter appearance. The bridges' V-shaped piers resemble open arms, enhancing the sense of connection between neighborhoods. When the entire Trinity River Vision project is completed, the Panther Island Bridges will become gateways, welcoming locals and visitors to a reimagined riverfront district.

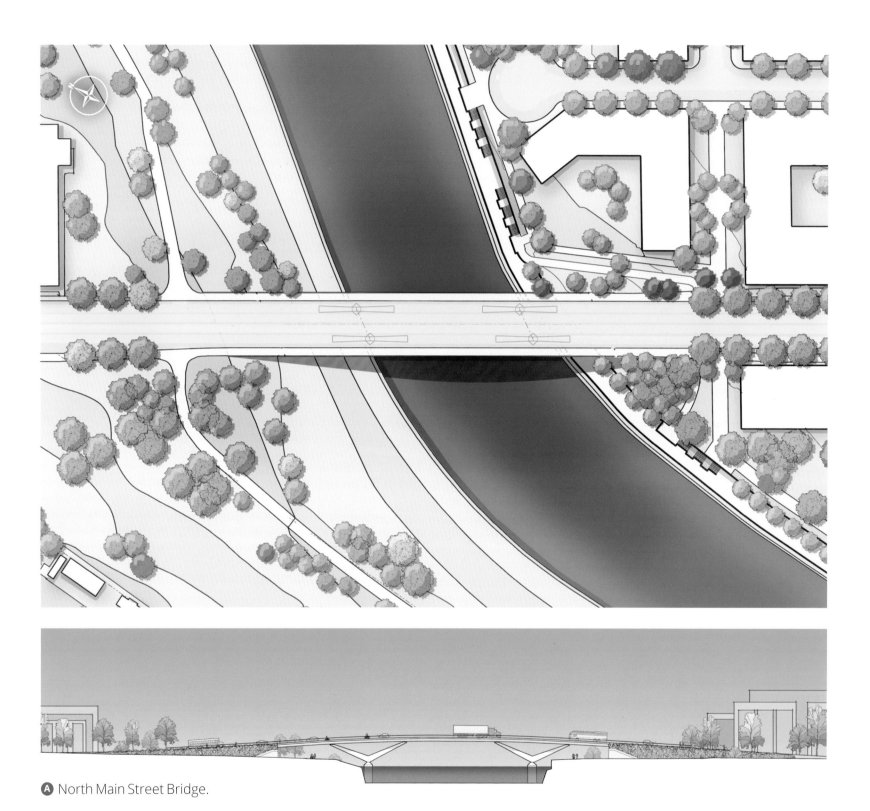

Ⓐ North Main Street Bridge.

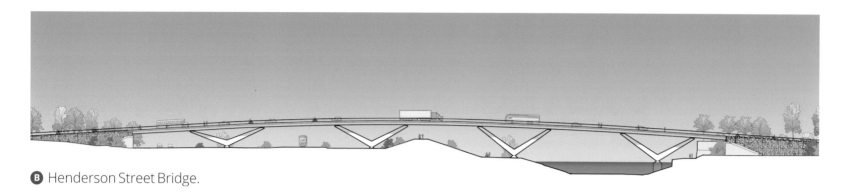

B Henderson Street Bridge.

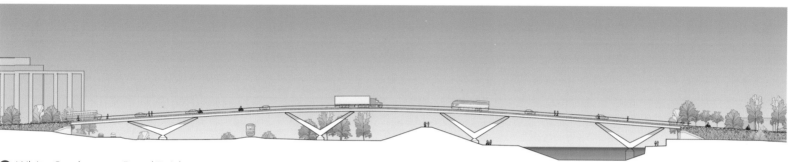

C White Settlement Road Bridge.

Trinity River Vision master plan showing the three bridges.

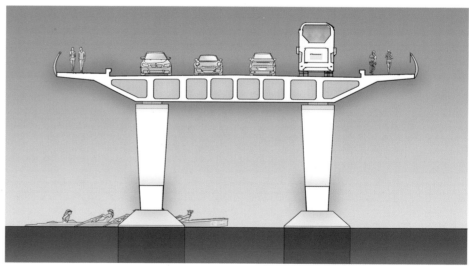

Bridge architectural features and detailing:

- Well-proportioned, slender V-shaped piers that resemble open arms.

- Arched pedestrian railings with razor-thin cables for unobstructed views and overall transparency.

- Elegantly and carefully detailed bridge abutments with diagonal rustications.

- Light-colored structural elements highlight the bridge along the new riverfront.

The three Panther Island Bridges are critical components of the Trinity River Vision project next to downtown Fort Worth, TX. The overall project will provide flood protection in addition to creating new river trails and parkland. This illustration shows how the river will be diverted under the three bridges in the future.

MOODY
bridge

MOODY
bridge

Location
Austin, TX

Completion date
2016

Width
10' (3 m)

Total length
300' (91.4 m)

Height of main towers
65' (19.8 m)

Vehicular clearance
15' (4.5 m)

The Moody Bridge is a striking inverted Fink truss bridge—the first of its kind in the United States. The bridge crosses over a well-traveled street at the University of Texas in Austin. The slender bridge provides a connection between academic buildings of the Moody College of Communication. An innovative approach was developed to avoid overloading the existing buildings while reducing the overall cost of the bridge. The effect is a simple, elegant network of towers and rods suspended above a busy urban street, removed enough to offer safe passage, yet placed and scaled to be part of the surrounding landscape and complement existing buildings.

The design and construction of the Moody Bridge posed the challenge of suspending a bridge between two existing buildings and coordinating construction on and above a busy street. The bridge design's single central tower foundation, with two long cantilevered arms, reduces the downward loads on the existing buildings, which otherwise would have required substantial modification. The design also eliminated the need to permanently reroute student pedestrian traffic around additional bridge supports closer to the buildings' entrances.

The aesthetic quality of the Moody Bridge derives from its simplicity and slenderness. For those looking at the bridge from the ground in daylight, it offers clean lines, an elegant composition of towers and rods, and a gateway to the large campus. In the evening the bridge offers a glowing silhouette. For students, faculty, and staff who cross between the buildings, it offers a defining view and a smooth path between academic buildings. The Moody Bridge achieves a synthesis of beauty, functionality, and efficient use of materials.

Design innovations and technological advancements at time of completion:

- First inverted Fink truss bridge in the United States.

- Efficient use of materials by sizing towers and rods to reflect the increasing loads toward the central tower foundation. Towers and rods are reduced in size and height when closer to the ends of the cantilevers for efficiency and economy.

- Main bridge foundation was built on a narrow street median that could not be widened.

Bridge architectural features and detailing:

- Towers, rods, and railing vertical supports all utilize tubular sections for visual consistency.

- Triangular shapes of the bridge relate to the profiles of existing campus buildings with pitched roofs.

- Bridge railings are illuminated with rectangular linear fixtures, creating a visual link between existing buildings.

- With its towers above the bridge deck, the bridge marks the location of the Moody College of Communication from a distance.

MARION STREET
bridge

MARION STREET
bridge

Location
Seattle, WA

Completion date
2024

Width
16' (4.8 m)

Total length
190' (57.9 m)

Main span length
110' (33.5 m)

Vehicular clearance
20' (6 m)

The Marion Street Bridge is located over the improved Alaskan Way along the waterfront in Seattle, Washington. The former elevated Alaskan Way viaduct, which carried State Route 99, was replaced with a tunnel under downtown. Since the removal of the double-deck viaduct, the waterfront is in the process of redevelopment, with new open spaces and landscaped areas along Elliott Bay. The bridge is an essential pedestrian connection to the city's largest multimodal terminal at Colman Dock. The crossing has become one of the busiest pedestrian bridges statewide.

The new concrete structure replaced a narrow, unattractive bridge that was not accessible. The enhanced connection has an exceptional width for a pedestrian bridge to allow for high peaks of usability during the morning and late afternoon commutes from Seattle's downtown to several residential neighborhoods across the bay.

The bridge design and overall streamlined appearance improves the image of the area and creates a gateway to the waterfront and beyond. The bridge's main span has a curved underside over the Alaskan Way and two floating cantilever arms. This allows the structure to be built independently from future buildings along the waterfront, providing maximum flexibility on construction timing. The main concrete reinforced V-shaped piers are sculptural, well-proportioned, and are illuminated at night. The bridge also features angled, curved pedestrian railings with integrated lighting, concealed utility lines, and a light-gray color. A consistent architectural design was also followed for the bridge approach to the main crossing. The Marion Street Bridge enhances and anchors the renaissance of Seattle's waterfront.

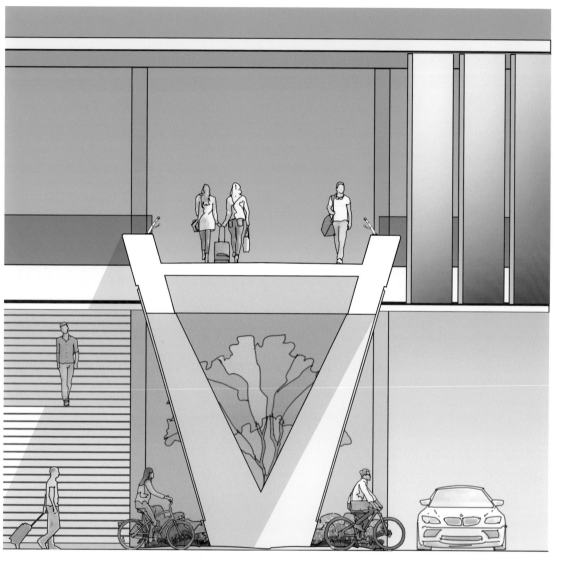

Bridge architectural features and detailing:

- Tapered V-shaped sculptural concrete piers.

- Aesthetic lighting enhances the bridge at night by highlighting the bridge profile from a distance.

- Use of architectural rustications to improve bridge proportions and visual slenderness.

- Consistent light-gray color of all bridge components to unify the overall appearance.

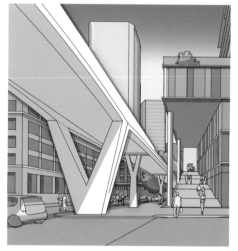

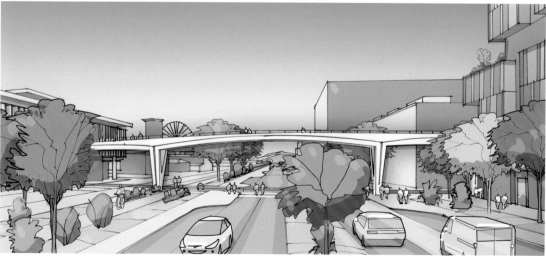

puente
CENTENARIO

puente
CENTENARIO

Location
Panama Canal,
Panama

Completion date
2004

Width
112' (34.1 m)

Total length
3,451' (1,051.8 m)

Main span length
1,380' (420.6 m)

Height of towers
600' (182.8 m)

Navigation clearance
262' (79.8 m)

The Puente Centenario is a slender cable-stayed bridge that carries six lanes of traffic and a median walkway across the Panama Canal. The landmark bridge is the second major crossing of the Panama Canal. It was built to help divert traffic away from the overcrowded four-lane Bridge of the Americas, close to downtown Panama City. The bridge is located 9.3 miles (15 km) north of the city and crosses the Culebra Cut near the Pedro Miguel canal locks, at one of the narrowest canal points.

The bridge became the main crossing of the Pan-American Highway while also relieving traffic congestion in the capital city. At its completion, the bridge had the longest main span for a concrete cable-stayed bridge in the Western Hemisphere. It was built in record time (less than two years) by using optimized construction methods. Two tall, slender towers support a single plane of cables anchored along the median between the traffic lanes. The towers are tapered with vertical architectural rustications and have a curved shape and a vertical reveal where the cables are anchored. The single plane of cables supports a single concrete segmental box with wide cantilevers.

The bridge is visible from great distances and contrasts dramatically with the lush rainforest setting. Aesthetic lighting illuminates the towers at night, making the bridge the sole focal point in an otherwise dark tropical jungle. The bridge commemorates the transit of the first ship through the canal in 1914. Designed throughout to appear modern and streamlined, the Puente Centenario is a source of civic pride for Panama and its people.

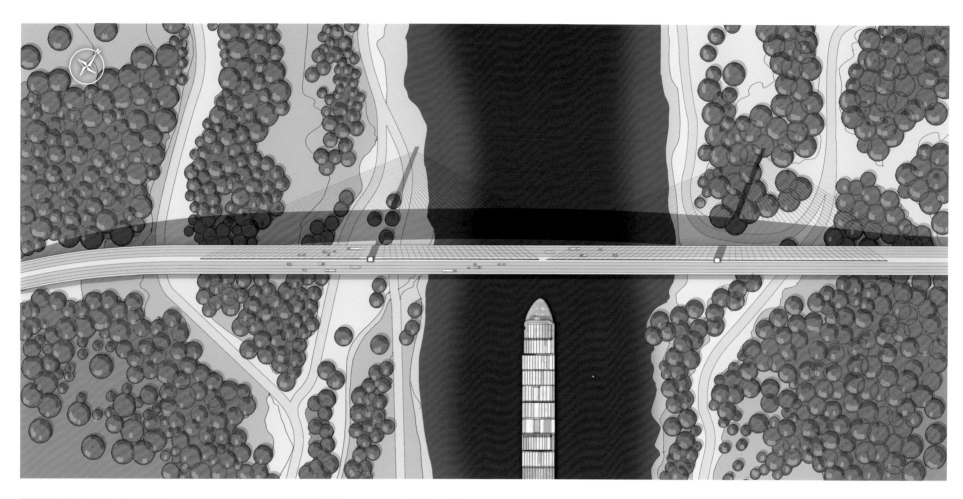

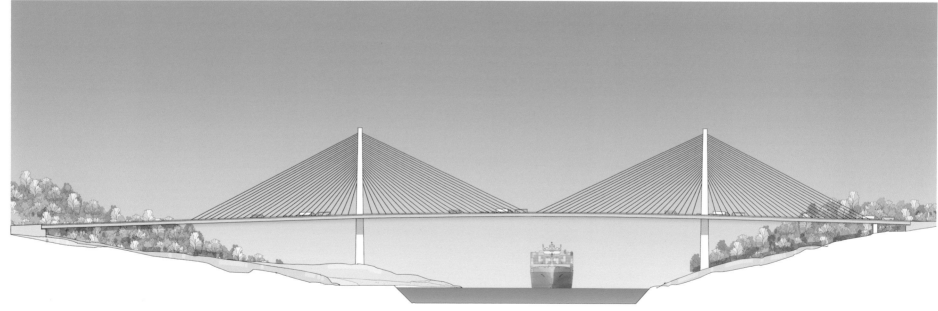

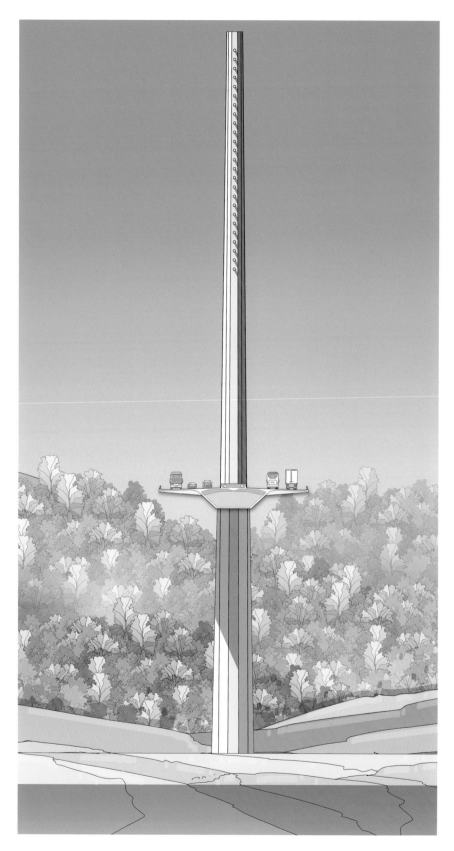

Bridge architectural features and detailing:

- Tapered, slender concrete towers with vertical rustications to improve proportions.

- Concrete box system with wide cantilevers that create shadow lines along the bridge facade, making the structure appear slender from a distance.

- Consistent materials, texture, and color for the entire bridge, including the superstructure and towers.

- Aesthetic lighting to enhance the crossing of the canal.

I-74 Bridge, Quad Cities, IA and IL.

BRIDGE
awards

ZAKIM BRIDGE

2003 National Council of Structural Engineers Associations (NCSEA) - Excellence in Structural Engineering Awards Outstanding Project Award (Multispan or Single Span over 150 ft.)

2003 International Bridge Conference - George S. Richardson Medal

2003 American Institute of Steel Construction (AISC) - National Steel Bridge Alliance Prize Bridge Award - Major Span

2004 American Society of Civil Engineers (ASCE) - Outstanding Civil Engineering Achievement Award

2004 American Council of Engineering Companies (ACEC) - Outstanding Civil Engineering Achievement Award

2004 American Society of Civil Engineers (ASCE) - Outstanding Projects and Leaders (OPAL) Award

2005 Boston Preservation Alliance - Preservation Achievement Award

LONGFELLOW BRIDGE

2019 National Trust for Historic Preservation - Richard H. Driehaus Foundation National Preservation Award

2019 American Public Works Association (APWA) Public Works Project of the Year Award – Historical Restoration/Preservation

2019 American Council of Engineering Companies (ACEC) - MA Engineering Excellence Award

2019 Engineering News-Record - New England Award of Merit - Highway Bridge

2019 Massachusetts Historical Commission Preservation Award

2019 Cambridge Historical Commission Preservation Award

2019 Boston Preservation Alliance - Preservation Achievement Award

2019 Preservation Massachusetts - Paul & Niki Tsongas Award

2020 International Bridge Conference - Abba G. Lichtenstein Medal

2020 Institute of Classical Architecture and Art - Bulfinch Award

2021 Traditional Building Magazine - Palladio Award

APPLETON BRIDGE

2019 International Bridge Conference - Arthur G. Hayden Medal

2019 The Chicago Athenaeum - American Architecture Award - Bridges and Infrastructure

2020 American Institute of Steel Construction - National Steel Bridge Alliance National Award - Special Purpose

2020 Rethinking the Future Awards - First Award - Transportation (Built)

2020   American Council of Engineering Companies
       (ACEC) - Engineering Excellence Awards (EEA)
       National Recognition Award for Exemplary
       Engineering Achievement

2021   Boston Preservation Alliance - Preservation
       Achievement Award

2021   Grands Prix du Design Platinum Award in
       Architecture/Infrastructure

MARKEY BRIDGE

2014   American Council of Engineering Companies
       of Massachusetts - Bronze Award - Honoring
       Outstanding Professional Design Excellence

2014   American Institute of Steel Construction -
       National Steel Bridge Alliance Prize Bridge Award
       - Special Purpose

FORE RIVER BRIDGE

2019   American Council of Engineering Companies
       (ACEC) - MA Engineering Excellence Award

WOODROW WILSON BRIDGE

2008   American Segmental Bridge Institute - Bridge
       Award of Excellence

2008   American Society of Civil Engineers (ASCE) -
       Outstanding Projects and Leaders (OPAL) Award

2008   America's Transportation Award Grand Prize
       Winner - American Association of State Highway
       and Transportation Officials (AASHTO)

2008   Mid-Atlantic Construction Best of 2008 Bridge -
       Award of Merit

2009   American Society of Civil Engineers/Maryland
       Section - Outstanding Civil Engineering
       Achievement for Major Construction Project

2009   American Institute of Steel Construction - National
       Steel Bridge Alliance Prize Bridge Award - Major
       Span

2009   American General Contractors (AGC) - Marvin M.
       Black Excellence in Partnering Award

COLUMBUS AIRPORT BRIDGES

2008   Portland Cement Association Bridge Award

LIBERTY BRIDGE

2005   American Institute of Steel Construction
       - National Steel Bridge Alliance Prize Bridge
       Award - Special Purpose

2005   American Galvanizers Association (AGA) -
       Excellence in Hot-Dip Galvanizing

2005   International Footbridge Award - Aesthetics
       Category for Medium Span Bridges

2005   International Bridge Conference - Arthur G.
       Hayden Award

2006   American Society of Civil Engineers (ASCE) -
       Outstanding Civil Engineering Achievement Award
       of Merit

2008   Waterfront Center - Excellence Annual Honor
        Award

2015   Rudy Bruner Award for Urban Excellence (RBA) -
        silver medal

I-74 MISSISSIPPI RIVER BRIDGE

2022   American Institute of Steel Construction -
        National Steel Bridge Alliance Prize Bridge Award
        - Major Span

2022   Eddy Award, River Action in the Quad Cities, IA
        and IL

2022   America's Transportation Award - Quality of Life/
        Community Development, Large Category

2023   Lincoln's Grand Conceptor Award - American
        Council of Civil Engineering of Illinois

2023   FIDIC - International Federation of Consulting
        Engineers, Geneva, Switzerland
        Highly Commended - Outstanding Project of the
        Year (Medium to Large)

2023   American Concrete Institute - First place in
        Excellence in Concrete Construction Awards,
        Infrastructure Category

GRIFFIN BRIDGE

2006   Federal Highway Administration Merit Award for
        Pedestrian Bridges

CARVER BRIDGE

2005   Outstanding Civil Engineering Project - American
        Society of Civil Engineers - Iowa Chapter

COMO PARK BRIDGE

2016   ACEC/MN Engineering Excellence Honor Award

2017   International Footbridge Award - Historic
        Renovation or Reuse

2017   Saint Paul Heritage Top Preservation Award

TILLEY BRIDGE

2013   American Council of Engineering - Engineering
        Excellence Award: National Recognition Awards -
        Texas, gold medal

2014   American Institute of Steel Construction
        - National Steel Bridge Alliance Prize Bridge
        Award - Special Purpose

2017   Waterfront Center - Excellence Annual
        Honor Award

MOODY BRIDGE

2016   Engineering News-Record - Small Project Award
        of Merit

PUENTE CENTENARIO

2005   American Segmental Bridge Institute (ASBI) -
        Bridge Award of Excellence

2007   Consulting Engineers and Land Surveyors of
        California (CELSOC) - Merit Award

Without the vision and support of our clients and collaborators, the bridges included in this book would not have become reality. My heartfelt gratitude goes to all of them.

Longfellow and Appleton Bridges, Boston, MA.

CLIENTS
+ collaborators

ZAKIM BRIDGE

Owner: Massachusetts Department of Transportation
Conceptual design: Christian Menn
Bridge architect/designer: Miguel Rosales
Preliminary designer: Bechtel/Parsons Brinckerhoff
Engineer of record: HNTB Corporation
Contractor: Atkinson Kiewit - joint venture

CHARLESTOWN BRIDGE

Owner: City of Boston, MA
Bridge architect/designer: Rosales + Partners
Engineer of record: Benesch
Contractor: J.F. White Contracting

LONGFELLOW BRIDGE

Owner: Massachusetts Department of Transportation
Bridge restoration architect and lighting designer:
Rosales + Partners
Preliminary design: Jacobs
Engineer of record: STV
Contractor: White/Skanska/Consigli - joint venture

APPLETON BRIDGE

Owner: Massachusetts Department of Conservation
and Recreation
Bridge architect/designer: Rosales + Partners
Preliminary design: Jacobs
Engineer of record: STV
Contractor: White/Skanska/Consigli - joint venture

NORTHERN AVENUE BRIDGE

Owner: City of Boston, MA
Bridge architect/designer: Rosales + Partners

MARKEY BRIDGE

Owner: City of Revere, MA
Bridge architect/designer: Rosales + Partners
Preliminary design: SBP
Engineer of record: AECOM
Contractor: Suffolk Construction Company

FORE RIVER BRIDGE

Owner: Massachusetts Department of Transportation
Bridge architect/designer: Rosales + Partners
Preliminary design engineer: STV
Engineer of record: Parson Corporation
Contractor: J.F. White Contracting/Skanska - joint venture

EAST 54TH STREET BRIDGE

Owner: City of New York, NY
Bridge architect/designer: Rosales + Partners
Preliminary design engineer: Stantec
Engineer of record: KC Engineering
Contractor: Skanska

WOODROW WILSON BRIDGE

Owner: Maryland Department of Transportation
Bridge architect/designer: Rosales + Partners
Engineer of record: Parson Corporation
Contractor: American Bridge/Edward Kraemer and Sons
- joint venture

COLUMBUS AIRPORT BRIDGES

Owner: Columbus Regional Airport Authority
Bridge architect/designer: Rosales + Partners
Engineer of record: RW Armstrong
Contractor: C.J. Mahan Construction Company

HICKORY RIVERWALK BRIDGE

Owner: City of Hickory, NC
Bridge architect/designer: Rosales + Partners
Engineer of record: Epstein
Contractor: W.C. English

LIBERTY BRIDGE

Owner: City of Greenville, SC
Bridge architect/designer: Rosales + Partners
Landscape architect: Arbor Engineering
Preliminary design: SBP
Engineer of record: Rosales + Partners
Contractor: Taylor and Murphy Construction Company

Left page, Moody Bridge, Austin, TX.

THROOP STREET BRIDGE

Owner: City of Chicago, IL
Bridge architect/designer: Rosales + Partners
Engineer of record: Benesch

RIVER NORTH BRIDGE

Owner: Oracle Corporation/City of Nashville, TN
Engineer/architect/designer: Rosales + Partners
Engineer of record: SBP

I-74 MISSISSIPPI RIVER BRIDGE

Owner: Iowa Department of Transportation
Bridge architect/designer: Rosales + Partners
Engineer of record: Modjeski and Masters
Contractor: Lunda Construction

HUNTSVILLE BRIDGE

Owner: City of Huntsville, AL
Engineer/architect/designer: Rosales + Partners

GRIFFIN BRIDGE

Owner: City of Des Moines, IA
Bridge architect/designer: Rosales + Partners
Engineer of record: HNTB Corporation
Contractor: Jensen/Cramer/United Contractors - joint venture

CARVER BRIDGE

Owner: City of Des Moines, IA
Bridge architect/designer: Rosales + Partners
Engineer of record: Earth Tech

BRUCE VENTO BRIDGE

Owner: City of Saint Paul, MN
Bridge architect/designer: Rosales + Partners
Landscape architect: Westwood Professional Services
Engineer of record: Rosales + Partners

COMO PARK BRIDGE

Owner: City of Saint Paul, MN
Bridge restoration architect and lighting designer:
Rosales + Partners
Landscape architect: Westwood Professional Services
Engineer of record: Clark Engineering
Contractor: LS Black Constructors

TILLEY BRIDGE

Owner: City of Fort Worth, TX
Bridge architect and lighting designer: Rosales + Partners
Preliminary design: SBP
Engineer of record: Freese and Nichols
Contractor: Rebcon

PANTHER ISLAND BRIDGES

Owner: City of Fort Worth, TX
Bridge architect/designer: Rosales + Partners
Engineer of record: Freese and Nichols
Contractor: Sterling Construction

MOODY BRIDGE

Owner: University of Texas at Austin
Bridge architect/designer: Rosales + Partners
Engineer of record: Freese and Nichols
Contractor: Flintco

MARION STREET BRIDGE

Owner: City of Seattle, WA
Bridge architect/designer: Rosales + Partners
Engineer of record: HDR
Contractor: Flatiron

PUENTE CENTENARIO

Owner: Canal de Panamá, Panamá
Bridge architect/designer: Rosales + Partners
Engineer of record: TY Lin International
Contractor: Bilfinger Berger AG

IMAGE credits

DUST JACKET · cover (Zakim Bridge) Juan Navarro · front flap Ian MacLellan · back cover (I-74 Mississippi River Bridge) Miller+Miller Architectural Photography · back flap (Appleton Bridge) David Desroches

HARD COVER · (Zakim Bridge) technical architectural drawing Fernando Algorta

INITIAL PAGES · p. 2 (Appleton Bridge) Juan Navarro · pp. 4–5 Alan Karchmer

BECOMING A BRIDGE DESIGNER · pp. 8, 12–13 Juan Navarro · p. 9 Ian MacLellan · p. 15 illustration Miguel Rosales · p. 16 Andy Ryan · p. 17 (top left) illustration Paul Stevenson Oles · p. 17 (top and bottom right) illustrations Randall Imai · pp. 17 (bottom left), 20 Peter Vanderwarker · pp. 18–19 illustration Dongik Lee · p. 21 Joel Benjamin · p. 22 John David Corey · p. 23 Kun Zhang · p. 24 Ekaterina Elagina/Deposit Photos · p. 25 Carolina Lau · pp. 28–31 technical architectural drawings Fernando Algorta

ZAKIM BRIDGE · pp. 32–33 Andy Ryan · p. 34 Greig Cranna · pp. 36–37 technical architectural drawings Fernando Algorta · p. 37 (top right) David Desroches · p. 38 Alan Karchmer · p. 39 (top left) Miguel Rosales · p. 39 (center and bottom left) John Woolf · pp. 39 (right), 40 (bottom left) Matt Conti · p. 40 (top left) Jon Bilous/Dreamstime · p. 40 (right) Elena Elisseeva/Dreamstime · p. 41 Robert Sansone

CHARLESTOWN BRIDGE · pp. 42–43, 44, 48, 49 photomontages and 3D renderings Uri Drachman · pp. 46–47 technical architectural drawings Fernando Algorta · p. 47 (center) David Desroches · p. 47 (bottom left and right) Jovan Tanasijevic

LONGFELLOW BRIDGE · pp. 50–51 Juan Navarro · pp. 52, 60 (bottom), 61 David Desroches · pp. 54–55, 57 (top) technical architectural drawings Fernando Algorta · pp. 56, 60 (top) John David Corey · p. 57 (center and bottom) Miguel Rosales · pp. 58, 59 (top row left and center, middle row right, bottom row right) Ian MacLellan · p. 59 (top row right, middle row left and center, bottom row left and center) Miguel Rosales · pp. 62–63 Alan Karchmer

APPLETON BRIDGE · pp. 64–65, 74–75 David Desroches · pp. 66, 69 (bottom right), 73 Alan Karchmer · pp. 68–69 technical architectural drawings and 3D sketches Fernando Algorta · pp. 70, 72 (all photos) Juan Navarro · p. 71 (all photos) Ian MacLellan

NORTHERN AVENUE BRIDGE · pp. 76–77, 78, 81 (right), 82–83 photomontages and renderings Uri Drachman · pp. 80, 81 technical architectural drawings and 3D sketches Fernando Algorta

MARKEY BRIDGE · pp. 84–91 (all photos) Carlos Arzaga · pp. 88–89 technical architectural drawings and 3D sketches Fernando Algorta

FORE RIVER BRIDGE · pp. 92–93, 97 (all photos) David Desroches · pp. 94, 98–99 Juan Navarro · p. 96 technical architectural drawings Fernando Algorta

EAST 54TH STREET BRIDGE · pp. 100–107 technical architectural drawings and 3D sketches Fernando Algorta · pp. 102, 105 (bottom), 106 (bottom left) Colin Miller

WOODROW WILSON BRIDGE · pp. 108–9 Christian Hinkle/Dreamstime · pp. 110, 117 Cameron Davidson · pp. 112–13 technical architectural drawings and 3D sketches Fernando Algorta · pp. 113 (bottom), 114–15, 116 (bottom) Alan Karchmer · p. 116 (top) Joseph Gruber/Alamy stock photo

COLUMBUS AIRPORT BRIDGES · pp. 118–25 (all photos) Brad Feinknopf · pp. 122–23 technical architectural drawings and 3D sketches Fernando Algorta

HICKORY RIVERWALK BRIDGE · pp. 126–27, 128 Alan Karchmer · pp. 130–31 technical architectural drawings and 3D sketches Fernando Algorta · p. 131 (bottom right) courtesy of City of Hickory, NC · pp. 131 (top right), 132 (all photos), 133 Jim Schmid

LIBERTY BRIDGE · pp. 134–35 Sean Pavone/Dreamstime · pp. 136, 140–41 Kevin Ruck/Dreamstime · pp. 138–39 technical architectural drawings Fernando Algorta · pp. 139 (right), 143 (all photos) Kun Zhang · p. 142 Mauritius Images GmbH/Alamy stock photo

THROOP STREET BRIDGE · pp. 144–51 3D renderings Uri Drachman · pp. 148–49 technical architectural drawings Fernando Algorta

RIVER NORTH BRIDGE · pp. 152–61 photomontages and 3D renderings Uri Drachman · pp. 156–57 technical architectural drawings Fernando Algorta

I-74 MISSISSIPPI RIVER BRIDGE · pp. 162–63, 169 (left) Gary Harris · pp. 164, 167 (top right), 168, 169 (top and bottom right) Miller+Miller Architectural Photography · pp. 166–67 technical architectural drawings Fernando Algorta · p. 167 (bottom right) David Desroches

HUNTSVILLE BRIDGE · pp. 170–77 photomontages Uri Drachman · pp. 174–75 technical architectural drawings Fernando Algorta

GRIFFIN BRIDGE · pp. 178–79, 184 (top), 185 Kevin Barber · pp. 182–83 technical architectural drawings Fernando Algorta · pp. 180, 183 (right), 184 (bottom) Kun Zhang

CARVER BRIDGE · pp. 186–87, 191 (bottom) Kevin Barber · p. 188 Jasey Michelle Bradwell · pp. 190–91 technical architectural drawings Fernando Algorta · pp. 192–93 Kun Zhang

BRUCE VENTO BRIDGE · pp. 194–201 photomontages Uri Drachman · pp. 198–99 technical architectural drawings Fernando Algorta

COMO PARK BRIDGE · all photos Dana Wheelock except: p. 208 (top left and right, bottom right) Nathan Holth · p. 208 (bottom left) Miguel Rosales · pp. 206–7 technical architectural drawings Fernando Algorta

TILLEY BRIDGE · pp. 210–11, 212, 215 (bottom left), 218 (bottom left and right) Greg Folkins · pp. 214–15 technical architectural drawings Fernando Algorta · p. 215 (right) Chad M. Davis, AIA · pp. 216–17, 218 (top), 219 Alan Karchmer

PANTHER ISLAND BRIDGES · pp. 220–21 Alan Karchmer · pp. 224–25, 227 technical architectural drawings and 3D sketches Fernando Algorta · pp. 222–26 (all photos) Greg Folkins

MOODY BRIDGE · all photos Justin Wallace except: pp. 232–33 technical architectural drawings Fernando Algorta · pp. 235 (bottom right), 236–37 (all photos) Alan Karchmer

MARION STREET BRIDGE · all photos Patrick Brennan · pp. 242–43, 244 (bottom) technical architectural drawings and 3D sketches Fernando Algorta

PUENTE CENTENARIO · pp. 246–47, 254–55 Danny Lehman · p. 248 Alfredo Maiquez/Dreamstime · pp. 250–51 technical architectural drawings Fernando Algorta · pp. 251 (right), 252 (left) José Ángel Murillo · p. 252 (top right) Gualberto Becerra/Dreamstime · p. 252 (bottom right) Rodolfo Aragundi (CC BY-SA2.0) · p. 253 Hemis/Alamy stock photo

FINAL PAGES · p. 256 Miller+Miller Architectural Photography · pp. 260–61 Juan Navarro · pp. 262, 266 Alan Karchmer · p. 268 Kun Zhang

Appleton Bridge, Boston, MA.

ACKNOWLEDGMENTS

Bridge design	Nicolas De Rycker
	Marcin Kasiak
	Jeffrey Plusen
Technical architectural drawings and 3D sketches	Fernando Algorta
Photomontages and 3D renderings	Uri Drachman
Advisor	Barbara Gomperts
Editor	Lissa Hanckel
Copy editor	Jennifer Harley
Graphic artist advisor	Olga Hazard
Final artwork	Carlos Andrino
Book graphic designer	Carolina Bran

www.oroeditions.com
info@oroeditions.com

Published by ORO Editions

Author: Miguel Rosales
Book Design: Carolina Bran
Project Manager: Jake Anderson

10 9 8 7 6 5 4 3 2 1 First Edition

ISBN: 978-1-961856-15-8

Prepress and print work by ORO Editions Inc.
Printed in China

ORO Editions makes a continuous effort to minimize the overall
carbon footprint of its publications. As part of this goal, ORO, in
association with Global ReLeaf, arranges to plant trees to replace
those used in the manufacturing of the paper produced for its books.
Global ReLeaf is an international campaign run by American Forests,
one of the world's oldest nonprofit conservation organizations. Global
ReLeaf is American Forests' education and action program that helps
individuals, organizations, agencies, and corporations improve the
local and global environment by planting and caring for trees.

Publisher's Cataloging-in-Publication
(Provided by Cassidy Cataloguing Services, Inc.).

Names: Rosales, Miguel, 1961- author.

Title: Bridges as structural art / Miguel Rosales.

Description: [Novato, California] : Oro Editions, [2024].

Identifiers: ISBN: 978-1-961856-15-8 | LCCN: 2024005376

Subjects: LCSH: Rosales, Miguel, |d 1961- --Aesthetics. | Rosales + Partners, Inc.
         | Bridges--Design and construction. | Bridges--Landscape architecture.
         | Bridges--Pictorial works. | Art and architecture.

Classification: LCC: TG140.R67 R67 2024 | DDC: 624/.092--dc23

Liberty Bridge, Greenville, SC.